W0043893

LABOR, EMPLOYMENT AND AGRICULTURAL DEVELOPMENT IN
WEST ASIA AND NORTH AFRICA

Labor, Employment and Agricultural Development in West Asia and North Africa

Edited by

DENNIS TULLY

The International Center for Agricultural Research in the Dry Areas (ICARDA), Aleppo, Syria

Kluwer Academic Publishers

DORDRECHT / BOSTON / LONDON

Library of Congress Cataloging-in-Publication Data

Labor, employment, and agricultural development in West Asia and North
  Africa / edited by Dennis Tully.
        p.   cm.
  Includes bibliographical references.
  ISBN-13:978-94-010-6773-7   e-ISBN-13:978-94-009-0635-8
  DOI:10.1007/978-94-009-0635-8

    1. Agriculture--Economic aspects--Middle East--Case studies.
  2. Agriculture--Economic aspects--Africa, North--Case studies.
  3. Agricultural laborers--Middle East--Case studies.
  4. Agricultural laborers--Africa, North--Case studies.
  HD2056.5.L33   1990
  338.1'0956--dc20   90-4702

ISBN-13:978-94-010-6773-7

Published by Kluwer Academic Publishers,
P.O. Box 17, 3300 AA Dordrecht, The Netherlands.

Kluwer Academic Publishers incorporates
the publishing programmes of
D. Reidel, Martinus Nijhoff, Dr W. Junk and MTP Press.

Sold and distributed in the U.S.A. and Canada
by Kluwer Academic Publishers,
101 Philip Drive, Norwell, MA 02061, U.S.A.

In all other countries, sold and distributed
by Kluwer Academic Publishers Group,
P.O. Box 322, 3300 AH Dordrecht, The Netherlands.

Printed on acid-free paper

All Rights Reserved
© 1990 ICARDA
Softcover reprint of the hardcover 1st edition 1990

No part of the material protected by this copyright notice may be reproduced or utilized i
any form or by any means, electronic or mechanical, including photocopying, recording or b
any information storage and retrieval system, without written permission from the copyrigh
owner.

Table of Contents

vi

Foreword

The basic objective of agricultural research at ICARDA is to enhance producer and consumer welfare through increasing the productivity, stability, and profitability of agriculture. Improved practices must be technically, economically, and socially suitable to farmer conditions.

The rainfed areas of West Asia and North Africa have highly variable environmental conditions as well as complex social and economic structures. In recent years, the region has been experiencing major changes in the relative availabilities and costs of the classical factors of production: land, labor and capital. These changes have important implications for the design of new agricultural technology.

On the one hand, the availability of labor may be an important factor determining the acceptability of new technology. On the other, it is important to consider the impact that technology can have on rural employment.

To develop a better awareness of these issues and their relevance to technology development, ICARDA initiated a project on Agricultural Labor and Technological Change (ALTC). The first stage of the project was a review of existing information on these issues; this review was published as a book under the title *Labor and Rainfed Agriculture in West Asia and North Africa*.

The second stage of the ALTC project was collaboration with regional scientists in original research on these issues. Eight case studies were funded and are published here. The authors have made significant contributions to our understanding of the linkage between labor issues and agricultural technology, and we hope that these studies will stimulate others to follow up with additional research in this area.

ICARDA greatly appreciates the financial assistance of the Ford Foundation and the International Development Research Centre (IDRC) of Canada, which allowed us to support these researchers and the production of this publication of their findings.

Nasrat Fadda
Director General
ICARDA

Dennis Tully (ed.), Labor, Employment and Agricultural Development in West Asia and North Africa, vii.
© 1990 ICARDA.

Acknowledgements

This publication is the culmination of the efforts of many people over the course of four years. The researchers committed themselves to meet high standards and tight deadlines with limited resources. The Ford Foundation and The International Development Research Centre (IDRC) of Canada not only made the ALTC project possible financially; their field staff made real contributions of ideas and time. Many of my colleagues at ICARDA worked with the researchers and myself to help design the case studies and the overall project. Finally, this publication benefits greatly from the skilled editing of James Pruess and the artwork of La Neu.

The Editor

Acknowledgements

List of Contributors

AKDER, HALIS

Department of Economics
Middle East Technical University
Inonu Bulvari
Ankara
Turkey

BEN ACHOUR, ARBI

Department of Rural Sociology
University of Missouri
Columbia
Missouri 65211
USA

BENATYA, DRISS

Institut Agronomique et Vétérinaire Hassan II
Direction du Développement Rural
B.P. 6202
Rabat
Morocco

BOUAITA, AHMED

Centre de Recherches en Economie Appliquée
pour le Développement (CREAD)
Enceinte
Université de Bouzarea
Rue Djamal-Eddine El Afghani
Al Hammadi
Bouzarea
Algeria

CHAULET, CLAUDINE

Centre de Recherches en Economie Appliquée
pour le Développement (CREAD)
Enceinte
Université de Bouzarea
Rue Djamal-Eddine El Afghani
Al Hammadi
Bouzarea
Algeria

Dennis Tully (ed.), Labor, Employment and Agricultural Development in West Asia and North Africa, xi–xiv.
© 1990 ICARDA.

ECEVIT, MEHMET

Department of Sociology
Middle East Technical University
Inonu Bulvari
Ankara
Turkey

EKINCI, NAZIM

Department of Economics
Middle East Technical University
Inonu Bulvari
Ankara
Turkey

ERKUS, AHMET

Department of Agricultural Economics
Faculty of Agriculture
Ankara University
Ankara
Turkey

GANA, ALIA

Institut National de la Recherche
Agronomique de Tunisie
Laboratoire d'Economie et de Sociologie
Rurale (INRAT)
2080 Avenue de l'Independence
Ariana
Tunis
Tunisia

GURKAN, A. ARSLAN

Department of Economics
Middle East Technical University
Inonu Bulvari
Ankara
Turkey

KARABLIEH, EMAD K.

Department of Agricultural Economics and
Extension
University of Jordan
Amman
Jordan

KASNAKOGLU, HALUK

Department of Economics
Middle East Technical University
Inonu Bulvari
Ankara
Turkey

KHALDI, RAOUDHA

Institut National de la Recherche
Agronomique de Tunisie
Laboratoire d'Economie et de Sociologie
Rurale (INRAT)
2080 Avenue de l'Independence
Ariana
Tunis
Tunisia

KHROUZ, DRISS

Département des Sciences Economiques
Faculté de Droit
Université de Fes
B.P. 42A
Fes
Morocco

KIRAL, TANER

Department of Agricultural Economics
Faculty of Agriculture
Ankara University
Ankara
Turkey

MARGHI, MOHA

Ministry of Agriculture
B.P. 1534 - Atlas
Fes
Morocco

SALEM, MAHMOUD ALI

Department of Agricultural Economics and
Extension
University of Jordan
Amman
Jordan

SIRMAN, NUKHET

Department of Anthropology
Middle East Technical University
Inonu Bulvari
Ankara
Turkey

TATLIDIL, F. FUSUN

Department of Agricultural Economics
Faculty of Agriculture
Ankara University
Ankara
Turkey

TATLIDIL, HASAN

Department of Agricultural Economics
Faculty of Agriculture
Ankara University
Ankara
Turkey

TULLY, DENNIS

The International Center for Agricultural
Research in the Dry Areas (ICARDA)
P.O. Box 5466
Aleppo
Syria

ZAGDOUNI, LARBI

Institut Agronomique et Vétérinaire Hassan II
Direction du Développement Rural
B.P. 6202
Rabat
Morocco

Introduction

DENNIS TULLY

This volume is the second publication from ICARDA's project on Agricultural Labor and Technological Change (ALTC). As the background and general issues are discussed in detail in the companion volume, *Labor and Rainfed Agriculture in West Asia and North Africa* (Tully 1990), only a brief overview will be provided here. The primary purpose of this volume is to present the original research findings of eight teams from West Asia and North Africa who participated in the ALTC project.

In most countries of West Asia and North Africa, rainfed agriculture under arid or semiarid conditions is a major part of the national economy and is the basis of life for large numbers of people. Under rainfed conditions, the agricultural labor market functions in ways that differ markedly from those in irrigated or urban situations. Thus, to best understand agricultural labor and employment issues in rainfed areas of the region, we must focus on these areas to the exclusion of irrigated agricultural areas.

Among the features specific to rainfed agriculture, the most relevant to employment are the highly seasonal nature of labor requirements and the low productivity of farms under current technical practices. As a consequence, hired or family labor can only be usefully employed for limited times and at low wages, and thus work opportunities are often insufficient to provide a minimal standard of living. The response of many rural people has been permanent emigration, and among those who remain in agriculture, seasonal or temporary migration or off-farm work are common. The fact that the first and most widely introduced technologies in rainfed areas have been labor replacing, particularly mechanization, has also contributed to this pattern.

Currently, interest is growing in the development and extension of technologies to increase the productivity of rainfed areas. This stems, on the one hand, from new awareness of the magnitude of potential production increases, as governments seek every opportunity to reduce imports of food. On the other hand, there is the social benefit to be gained by the concomitant increase in the productivity of labor, which can be expected to improve rural incomes and nutrition, and reduce the tide of migration to overcrowded urban areas.

The ALTC project was developed by ICARDA with input from scientists in many national programs. The purpose has been to generate information on labor issues so that scientists, policy makers, and development admini-

Dennis Tully (ed.), Labor, Employment and Agricultural Development in West Asia and North Africa, 1–5.
© 1990 ICARDA.

strators can make better choices. Better information should help us select technologies that are more appropriate to the labor available in rainfed areas, while having a positive impact on rural employment.

From its initiation, the ALTC project was designed to be exploratory and decentralized; there was no labor prescription written by ICARDA. Rather, the strategy was to seek out scientists in national programs with an interest in this topic, and learn what they perceived to be the problems, and their methods for arriving at solutions.

As a first step, review papers at the national level were commissioned for eight countries and, at the regional level, for four themes; these appear in the companion volume mentioned above (Tully 1990). These papers clearly demonstrated that the labor situation of the rainfed areas is similar across the region in many respects: the prevalence of migration and part-time farming, the aging of the rural labor force, problems of land fragmentation, etcetera. However, these similarities are modulated by greatly varying histories, policies, and resource endowments of the countries in the region. A regional pooling of information therefore appears to be a promising approach, as it will allow for comparison and interpretation of differences.

The review papers also revealed that in most countries, there has been very little study of these issues at the community level. Confusions and paradoxes in national data may well result from the aggregation of data from quite different local situations. The factors that cause a new agricultural technology to have a positive impact in one community and a negative impact in another often cannot be revealed by national data. Of course, it is not practical to study every community to which a new technology might be offered. However, the similarities found in the review papers offer hope that a reasonable number of community-level studies throughout the region will show which differences are significant and which are irrelevant to technological change. The research whose results are presented in this volume has been conducted in this spirit.

In addition to the differences among communities, these case studies illustrate the differences of approach and methodology which prevail in this region. The juxtaposition in this volume of the approaches of differing disciplines and epistemological schools is representative of the variation in the region, which could present an obstacle to interaction. However, after working with the authors over several years and observing them together at a workshop in 1988, I firmly believe that this diversity represents one more opportunity for fruitful interaction. The researchers found the differences stimulating and were very much in favor of continuing regionally coordinated efforts in applied social science research. Because of their practical orientation and shared commitment to improving the situation of poor farmers, they have been able to easily overcome academic differences.

The papers in this volume are a rich source of data. While the weight given to the several issues varies from one paper to another, there are a number of recurring themes. To the extent that the papers are considered together, they complement each other in describing a continuum of situations, rather than eight isolated cases.

One factor that shows its importance in one form or another in all cases is labor productivity, which is based in large part on the underlying productivity of agriculture. Khrouz and Marghi's paper about a relatively rainy and fertile area of Morocco provides the most optimistic scenario. In this case, the introduction of new technologies, some of which are labor saving and some of which require additional labor, has resulted in aggregate not only in more employment, but in more productive employment with higher incomes. In addition, the new crop diversification has reduced the seasonality of labor, offsetting the opposite impact of mechanization. Similar results are found in the papers by Gana and Khaldi, and Erkus et al.

Although in most study areas rainfall is insufficient to support the range of new technologies and crops found in this particular district in Morocco, the case study does describe one possible model for the introduction of new technologies. At the other extreme, one might place cases where the potential of agriculture in the environment is very low. Such conditions seem to prevail in the dry areas described by Zagdouni and Benatya, and also by Ben Achour. In these areas, new crops are least likely to be successful and the balance of new technology appears to be labor replacing. As such it enhances the productivity of the remaining labor but makes emigration, off-farm labor, and part-time farming inevitable for much of the population.

In terms of farmer strategies, small farm size appears to have much the same effect as aridity. On small farms the gains associated with new technologies would appear to be so small in absolute terms that they contribute little to farm incomes, and as such do not significantly change labor productivity. Thus whether farm productivity is low because of small farm size or low environmental potential, the response is likely to be the same. From the farmer's point of view, the best way to increase the productivity of family labor is through off-farm activities. (Increased livestock activities may also play a role, as in the case presented by Bouaita and Chaulet). To be attractive in these situations, new technologies that require substantial labor will have to be much more productive than those that do not affect current low levels of labor input.

The fact is that returns to labor in rural areas, whether we are talking about wages paid to laborers or returns to family labor, are judged by comparison with urban or off-farm wages in jobs of comparable skill level. If rural compensation is not attractive, due to low productivity or low commodity prices, farms will be neglected and farmers will depend on off-farm income. The most extreme case presented of this type is from Jordan (Karablieh and Salem). In this case, farmers are better off emigrating and hiring labor to replace themselves on the farms.

The complexity of the mechanization issue is also demonstrated by these case studies. While restricted to large farms in some cases, in others the ownership of machinery by medium-sized or small farms serves the same function as off-farm labor, in that the owners can then obtain a good income by providing custom services to other farmers. Agricultural machinery also brings up questions of rural-urban linkages, particularly with respect to capital flows. Ownership may be limited to large landowners by financing policies, but where this is not the case machinery may strengthen urban-rural financial as well as familial ties. An interesting case is described by Kasnakoglu et al., in which a father paid to educate three sons who attained urban employment; they loaned their father money to buy a tractor, and he repays the debt by sending them money each year out of the harvest.

Another factor suggested by several studies is the linkage between mechanization, fragmentation, and sharecropping. As farmers with small or medium-sized holdings purchase machinery, they seek to expand the size of their holdings in order to make full use of it. This is leading to a new form of sharecropping in which smallholders lease land among themselves, depending on their overall strategy; one buys equipment and tries to make it worthwhile in the countryside, while his neighbor goes to the city and leases his land for a share of the harvest or, sometimes, for rent. The prevalence of these patterns makes it clear that many emigrants wish to preserve the option of coming back to the land.

The local-level focus of these studies brings out the importance of farmer choices in these processes. Mechanization and off-farm employment are not just broad trends that are occurring of their own accord; they represent choices that farmers make to maximize their household incomes (not necessarily farm incomes). The use by some authors of a typological approach with detailed fieldwork helps to identify the source of some puzzles in aggregate statistics. For example, why do the smallest farms with the highest ratios of labor to land specialize in low-labor winter cereals and use mechanical techniques at the same level as the largest farms, while medium-sized farms use the most labor-intensive techniques? In the case studied by Zagdouni and Benatya, this situation reflects three different strategies. Large holdings are short of labor; small holdings depend on wage labor; but medium-sized holdings can effectively use labor on their farms. Gana and Khaldi use a similar typological approach.

These papers are intended to stimulate an exchange of information across national boundaries and between scientists of different disciplines. There is still no prescription to be written for improved technology that will have optimal effects on employment, productivity, and income. With the authors, I hope that by providing a better picture of the situation in which farmers find themselves, we will allow more scientists and administrators to put themselves in the farmers' position when they are thinking about resource allocation. The situation of the arid zones is not comfortable for anyone at present. Farmers are ready to grasp at any reasonable approach to improving their productivity. But to be reasonable means appreciating not only the constraints but also the alternative opportunities that farmers face. When technologies are identified that improve farm productivity and also fit with overall household strategy, they will be enthusiastically received by farmers.

Reference

Tully, Dennis (ed.) 1990. *Labor and Rainfed Agriculture in West Asia and North Africa.* Dordrecht: Kluwer Academic Publishers.

The Impact of Technology on Employment in the Rainfed Farming Areas of Irbid District, Jordan

EMAD K. KARABLIEH and MAHMOUD ALI SALEM

Introduction

Since the early 1960s, there has been a growing emphasis on the rapid development and diffusion of agricultural technology in developing countries. Some countries have experienced the 'Green Revolution,' the adoption of new high-yield varieties of wheat, rice, and other cereals. The acceleration in population growth in low-income developing nations and continued growth in per capita income have contributed to greatly increased demand for food and agricultural products. This demand, recent world food scarcities, and high food prices emphasize the need to increase agricultural productivity in developing countries, including Jordan.

There is a great potential for increasing the production of wheat and legumes in the rainfed areas of Jordan through the use of better technology. However, both the adoption rate of technology and the growth rate of productivity are low. This low productivity in the rainfed sector is due to the use of land unsuitable for cultivation and to irregular rainfall that causes low yields and induces farmers to minimize their risks by limiting inputs to seed and plowing.

Although farmers need technology in the rainfed areas of Jordan to increase agricultural productivity, the adoption process is not that simple. It requires an awareness of constraints that local farmers experience and the participation of the farmers themselves. There are a number of constraints on technology adoption in Jordan.

In many cases, the available technology is not appropriate for the specific agroclimatic conditions that farmers experience, or farmers do not understand how the technology works. Farm size, type of tenure, or topography can be constraints on the adoption of new mechanized technology, such as tractors.

In addition, the availability of labor and capital can influence the adoption of technology. Many studies support the notion that technology for high-yield varieties (HYV) requires more labor (Ruttan 1977). The availability of family labor or the shortage of labor in general can affect the adoption of technology.

Access to capital is often an important factor in explaining variation in the rate of technology adoption. This is especially true in the case of mechanization, which requires a larger initial investment than HYV technology.

Dennis Tully (ed.), Labor, Employment and Agricultural Development in West Asia and North Africa, 7–30.
© 1990 ICARDA.

Agricultural prices, access to markets, and supplies of high quality seed may have influenced the adoption of technology in a number of developing countries. Risk and uncertainty seem to contribute to differential adoption rates among farmers. In general, low-income farmers are considered averse to risk because they are poor and lack the capital to withstand wide variations in expenditure and income.

Although many factors affect the adoption of technology in the rainfed farming areas of Jordan, the overriding one seems to be economics. Farmers have always accepted those new ideas and practices that increase net returns, rejecting those that do not. They are well aware of the agronomic benefits of new technology. However, they tend to adopt only those innovations they find economical. The adoption of appropriate technology would have a positive impact on output, employment, economic growth, and income distribution.

The impact of technology on employment in many countries has been widely recognized and increasingly documented. However, there is a general consensus that technology adoption has had positive impacts on employment in some developing countries and negative impacts in others. Institutional and policy environments have influenced the final effects of technology on employment. However, the implications for rainfed farming in the Middle East are not clear and require further study.

In the Middle East, women have an important role in farming. Rassam (1984) examined the relative contribution of men, women, and children to agricultural production in the northwestern Syrian rural household, focusing primarily on women's labor and the impact of technology on their labor. The results of the study indicated that the average contribution of women to total household production somewhat exceeds that of men. They provide more labor than men for cereal and legume crops, although men provide more labor in tree and summer crops. The findings of the study showed that new technologies would have diverse effects on women's work in general.

The results of further analysis (Rassam and Tully 1986) indicated that the hired labor for mechanical operations was predominantly male, while that for manual operations was mostly female. In addition, the ongoing mechanization of production would continue to reduce women's agricultural activities, both in the household and as hired labor.

Richards and Martin (1983) found that in Egypt's Nile delta region, 30% of the field crop laborers are women. Furthermore, social constraints on women's participation in agricultural labor combined with increasing male education and emigration are contributing to the current 'labor shortage' crisis.

As we will show later, women also contribute to the labor requirements of rainfed farming in Jordan.

Problems in Jordanian Agriculture

Jordan's agricultural sector is unable to satisfy the country's need for various food and agricultural products. The contribution of agriculture to GNP has declined from 20% in the early 1960s to 7.2% in 1986 (DOS 1986).

Jordan depends on imports to meet its consumption requirements for a large number of agricultural commodities. Imports of agricultural commodities satisfied more than 50% of domestic needs for grain and meat. The average annual imports of agricultural commodities amounted to nearly 35 million dinars (JOD) in 1986 (1 USD = 0.30 JOD) while exports of agricultural commodities were only 12 million JOD (DOS 1986).

In regard to the land available for agriculture, and the use of this land, Jordan is faced with problems and constraints that affect agricultural production and growth. The land area of Jordan is about 9.3 million hectares but the total cultivable area is only about 500,000 hectares of which 93% is rainfed and 7% (about 36,000 hectares) is completely irrigated. A little over 91% of the land area is in the dry desert region, which has no economic importance for Jordan except for occasional sporadic grazing (Table 1).

In addition to natural constraints, there are serious problems with land tenure. Sixteen percent of Jordan's farms are smaller than half a hectare, 64% are smaller than 5 hectares, only 2% of the farms are between 50 and 500 hectares, and only 0.03% of them are over 500 hectares in size (DOS 1983).

In addition, land holdings in Jordan are small and fragmented, with holdings divided into an average of 3.5 fragments. Thus, a farmer with three hectares may own three pieces of land in three different places. Most farmland in Jordan is owner-operated, and about 16% of it is rented. Urbanization and land speculation have further reduced Jordan's small high-potential agricultural land resource. (See Table 2 for the distribution of holdings by district.)

These problems have led to an increase in the number of part-time farmers (Arabiat and Snober 1984). In addition, younger people are less interested in farming as a profession, and many farmers have been forced to emigrate or change occupation (Barham 1986).

The main objective of this case study is to examine the impact of new technology on employment in the rainfed farming areas of Irbid district. Specifically, we will consider the current situation with respect to employment

Table 1. Climatological Zones in Jordan.

Zone	Average Rainfall	Area (thousand ha)	Percentage
Arid desert	100 mm	7500	81.1
Desert	100–200 mm	960	10.3
Marginal	200–300 mm	530	5.7
Semiarid	300–400 mm	170	1.8
Semihumid	500 mm	100	1.1
Total		9260	100.0

Source: HKJ (1980).

Table 2. Distribution of Holdings by District.

District	Number of Holdings	Area (ha)	Percentage
Amman	7,965	98,100	15.6
Irbid	29,915	196,700	58.9
Balqa	546	22,900	10.8
Karak	589	56,000	11.6
Ma'an	154	16,700	3.0
Total	50,790	390,400	100.0

Source: DOS (1982).

and technology, the relative labor contribution of men, women, and children in the cultivation of cereals and legumes, and the socioeconomic factors that affect labor input and explain the adoption of technology.

Employment and Agricultural Technology in Jordan

Agriculture in Jordan is heavily mechanized. Most farmers obtain their machinery through private rental services. Farm mechanization in the rainfed areas of Jordan is limited to plowing, planting, and harvesting grain.

This trend toward agricultural mechanization in Jordan is the result of labor shortages and rising labor costs. Many Jordanian workers have migrated from the rural areas to the oil-rich Arab states. This migration has caused shortages during periods of peak labor demand. These shortages increase the costs of labor for harvesting crops by hand, particularly lentils (Duwayri 1985).

Jordan's emigrant workers in the Arab states are about 38% of its total national work force. In 1968, about 40% of the labor force was engaged in agriculture; this percentage decreased to 7.8% in 1985 (Khasawneh 1986).

There is some disagreement on the effects of mechanized production techniques on employment. Some authors (e.g., Pinstrup-Anderson 1979) state that agricultural employment has increased due to the fact that either larger areas are cultivated or more intensive cultivation is taking place. A second group (e.g., Lipton 1977) states that employment has decreased in the agricultural sector, but there is increased employment in other related sectors. A third group of authors (e.g., Sulkharomana 1983) admits that mechanization decreases employment opportunities but considers this a short-term disadvantage. A fourth group (e.g., Begum 1985) claims a reduction in employment, with a severe degradation of living standards for the poorest members of society.

The use of fertilizer is new in Jordan. While farmers in irrigated areas apply fertilizer to vegetables and fruit trees, the use of fertilizer in the rainfed areas is still low.

Arabiat, Nygaard, and Somel (1982) commented on the slow adoption rate, pointing out that 50% of the wheat crop is used for home consumption, and farmers are not motivated to adopt improved technologies, such as fertilizer.

In addition, agroclimàtic conditions in Jordan, especially the variability of rainfall from year to year, introduce a high uncertainty and risk in agricultural production. According to the survey by Arabiat and his colleagues, soil conditions and the risk associated with rainfall are the most important factors that make the farmer avoid fertilizer application.

The Land Tenure System in Jordan

The equal distribution of a father's farmland among his sons at death has contributed over the years to a severe fragmentation of land and a reduction in the size of holdings. The small size of plots, combined with low income per unit area, discourages mechanization in rainfed areas. Furthermore, the use of mechanized equipment may be time-consuming if a farmer's tiny plots are scattered over a wide area.

The Agricultural Labor Force

The Jordanian labor force has played a major role in the economic development of Jordan. Prior to the 1970s, the government supported the outflow of Jordanian manpower for work abroad because the labor market was somewhat loose and there was a surplus of labor. The first comprehensive development plan (1973-75) set as one of its major targets the reduction of the rate of unemployment, which was estimated at 8% (Hourani 1985), and the creation of new job opportunities due to the dramatic rise in oil prices after 1974.

Nevertheless, in the beginning of 1984, the unemployment rate was once again 8% of the total labor force. This unemployment was due to the increase in the number of graduates and the return of the labor force from abroad. The agricultural labor market suffered from a shortage of Jordanian labor. Because of the availability of relatively cheap migrant labor, many Jordanian farmers have become more interested in working as sharecroppers or renting land. It is estimated that 14,562 migrant agricultural laborers were working in Jordan in 1986. A survey conducted by the Jordan Valley Authority showed that 95% were Egyptians and the rest Pakistanis. Interviewed workers reported that they worked nine months on the average before returning home. Their wages ranged from 2 to 2.5 JOD per day. Laborers usually worked from six in the morning until noon. In addition to cash wages, employers also provided accommodations together with free meals, usually breakfast for labor hired by the day, and three meals for workers who were paid monthly. The interviewed workers also reported that more than three quarters of the Egyptian workers (76%) were illiterate (Hourani 1985).

Agricultural Employment in the Rainfed Areas

Because of their traditional farming practices, Jordan's rainfed farmers are described as averse to taking risks with agricultural innovations. Nevertheless, Jordan's urban areas and other Arab countries have lured away potential farm labor, and farmers have increased their use of labor-saving equipment, primarily tractors and combines, in the cultivation of rainfed grain crops. A survey conducted by Arabiat and Snober (1984) revealed that 77.4% of the farmers hired tractor operators. In addition, the supply of other equipment, such as boom sprayers, grain drills, orchard sprayers, and harrows, was not sufficient.

Jordanian labor has moved from the agricultural sector to urban areas or abroad, or become involved in the daily migration from village to city. Because most farmers are part-time agriculturalists, they usually hold jobs in government or the private sector. In recent years, farmers have become less dependent on their family members for agricultural operations.

Methodology

The Department of Statistics provided a list of all villages in Irbid district, including number of households and amount of land in each village. We selected a stratified random sample of farm households, based on size of farm, levels of technology (mechanization, fertilizers, pesticides, new methods of plowing), and field crops (cereals and legumes). We interviewed fifty farmers in three villages of the district: Nu'aima, located in Zone 1 (200-300 mm annual rainfall), and the villages of Husun and Al-Shajara, located in Zone 2 (300-400 mm annual rainfall).

By means of a structured questionnaire, we obtained data from the farmer and sometimes from both the farmer and the farmer's wife simultaneously. Sons of the farmers also took part in the interview sessions. The questionnaire made distinctions between household labor (age and sex) and hired labor (locally hired and migrant workers). The number of workers and the number of days as well as the number of hours spent in each agricultural operation were calculated for both cereals and legumes. We used number of hours to measure the contribution of the various labor groups to agricultural production activities.

In addition, the questionnaire elicited information dealing with subjects such as family size, land tenure, hired labor, farm machinery, institutional linkages, animal husbandry, and family income.

Characteristics of Sample Farms

The interviewed farmers, on the average, were old (59 years) and had large families (11.5 persons). Families had an average number of 5.9 males and

Table 3. Percentage of Income Generated from Different Activities.

Item	Male	Female	Total
Farm income	51	2	53
Livestock	1	1	2
Agricultural (off-farm)	6	0	6
Nonagricultural	37	2	39
Total	95	5	100

5.6 females. The average number of students in each family was 3.9, and the average number of absentees from the village was 2.6. Farmers were quite experienced in farming, with an average of 35 years of experience. We found that 18% of the farmers were illiterate and only 18% of them completed secondary school. The area under cultivation averaged 27.9 hectares and ranged from 3 to 165. The average land holding was 35.8 hectares, with 16% of the farmers owning less than 5 hectares, 38% less than 10 hectares, 82% less than 50 hectares, and 8% owning more than 100 hectares.

About 96% of the farmers planted wheat, with an average of 24.6 hectares of wheat in an average of 3.5 parcels. They also planted an average of 14.5 hectares of barley. Sixty-six percent of the farmers planted legumes, with an average of 7 hectares in an average of 1.5 parcels. Twenty percent of the farmers belonged to agricultural cooperatives, 64% considered themselves to be in good health, and 58% of them were full-time farmers.

The income generated from farm activities constituted 53% of the total family income compared to 39% from nonagricultural activities. In Table 3, we show the percentages of income generated from agricultural and nonagricultural activities by gender. Agricultural (off-farm) income is obtained by the owners of tractors and combines who hire out their services to other farmers.

Crop rotation practices in Irbid district vary from one area to another according to agricultural zone. In Zone 1 (200–300 mm annual rainfall), farmers observe a two-year crop rotation: wheat or barley alternating with fallow. The wheat-fallow rotation occupies about 80% of the area in this zone. In Zone 2, the rainfall pattern is more reliable (300–400 mm annually). A three-year crop rotation occurs most often, mainly wheat-legumes-fallow or summer crops. Only 2% of the farmers planted wheat continuously.

Demographic Characteristics

The percentages of males and females were approximately equal in the sample. Males under 20 years of age formed 19.2% of the sample, and females under

Table 4. Level and Percentage of Education in Sample Households.

Educational Characteristics	Percentage
Illiterate	10.60
Literate (can read and write but don't have certificate)	1.87
Elementary school	23.70
Preparatory school	19.33
Secondary school	23.07
College	6.23
University	6.23
Graduate studies	2.07
Age less than 6 years	6.86
Total	100.00

20 were 19.9% of the sample. The older population, 60 years of age and over, made up 12.9% of the sample.

In Table 4, we include information on the level of education among members of the sample households.

Results of Multiple Regression Analysis

We analyzed the data to determine the factors affecting male and female labor input, and the relative importance of hired and household labor.

We used a multiple regression model to statistically test the relationships between labor input (number of hours by group) and a number of socioeconomic variables. These included the amount of land owned, farmed, and planted, characteristics of the farmers (age, education, experience in farm management, health status), demographic characteristics of farm families (number of males and females, family size, number of students, number of absent family members), source of family income, number of off-farm work days, membership in agricultural societies, attitudes toward risk, and sites of villages.

We used stepwise regression at the 5% level of significance to consider the importance of variables. In Table 5, we present the results in regard to labor input.

These results show that area planted in crops was a statistically significant factor for most of the different categories of labor. An increase in area increased the need for labor. The labor hours in Al-Shajara were greater than in Nu'aima and Husun, because of the three-year crop rotation (wheat-legumes-summer crops). An increase in the size of holding corresponded to a decrease in labor input because of the efficiency in the utilization of hired labor. An increase in the number of males in the family increased labor input. Families with

Table 5. Multiple Regression Coefficients for Independent Variables Related to Labor Input (Hours).

Variables Entered in Equation	Household			Local Hired			Migrant	Total Hired	Total (Household & Hired)
	Male	Female	Total	Male	Female	Total			
Farmer's age	–	–	–	–	10.7	10.7	–	–	–
Years of education	–	–	–	-21.2	–	–	–	–	–
Years farming	–	–	–	–	-5.9	–	-10.2	–	–
Area farmed (ha)	-4.9	–	-6.9	–	–	–	–	–	7.9
Area owned (ha)	13.9	–	16.4	9.6	–	11.1	–	6.2	29
Area planted (ha)	46.5	–	–	–	-23.6	–	–	–	–
Family size	–	–	80	65.7	–	–	–	–	138
Number of males	–	–	–	–	–	–	–	–	–
Number of females	–	–	–	–	–	–	–	–	–
Number of students	-50.5	–	-60	-44.3	–	-33.9	-58.7	-89	-176
Number of absentees	-54.4	–	-63	–	–	-104	–	-123	-220
Off-farm work (days/yr)	–	–	–	–	–	–	–	–	–
Farm income (% of total)	–	-.9	–	–	–	–	-3.2	–	–
Off-farm income (% of total)	–	–	–	–	–	–	–	–	–
Risk index[a]	–	–	–	–	–	–	–	–	–
Member of agricultural coop[b]	–	–	–	–	–	–	–	–	–
Farmer healthy[b]	–	–	–	–	112	–	–	–	–
Planted after rain[b]	-107	-57	-146	–	–	–	-434	-382	-544
Village[c]	–	–	–	–	–	–	–	211.7	296
Intercept	-47	122	145	-64	-186	-518	1419	1036	300
R²	.61	0.1	.55	.57	.22	.44	.28	.34	.55
F ratio	11.62	2.86	8.98	9.67	3.25	7.02	4.5	4.6	7.44

[a] kg per ha in bad years/kg per ha in normal years divided by probability of bad years/probability of normal years.

[b] Dummy variables, with 1 if positive, 0 if negative.

[c] 1=Husun, 2=Nu'aima, 3=Al-Shajara

– not significant at p=.05

Table 6. Multiple Regression Coefficients for Independent Variables Related to Technology Adoption.

Variables Entered in Equation	Fertilizer Applied (kg/ha)	Herbicide
Farmer labor hours	.014	–
Hired labor hours	–	–
Migrant labor hours	–	.0002
Farmer's age	–	–.01
Member of agricultural coop[b]	–	–.26
Risk index[a]	–25.7	–.34
Years farming	–	–
Years of education	–	–
Area planted (ha)	.061	–
Farmer healthy[b]	–	–
Farm income (% of total)	–	–
Village[c]	–	–
Intercept	87.6	1.27
R^2	0.26	0.21
F ratio	5.4	3.16

[a] kg per ha in bad years/kg per ha in normal years divided by probability of bad years/probability of normal years.
[b] Dummy variables, with 1 if positive, 0 if negative.
[c] 1=Husun, 2=Nu'aima, 3=Al-Shajara
– not significant at p=.05

more students and absent family members devoted fewer hours to agriculture.

We used a second multiple regression model to statistically test the factors that affect a farmer's adoption of technology. In Table 6, we present the results of the analysis of factors that affect the use of fertilizer and herbicide. Only thirteen of the nineteen variables were used in this multiple regression linear analysis.

Labor Input in On-Farm Activities

Farm labor was the primary constraint affecting agricultural production. Both family labor and hired labor were important factors that determined the supply and demand for labor in conjunction with the introduction of new techniques of agricultural production.

In collecting these data, we considered individuals fifteen years old and above to be adults; children under eight years of age had no importance in economic activities. In the following tables, we present data on the labor contributions of different categories of workers, in terms of number of workers, number of work days, and number of hours per day for each crop and each task. We also used the number of hours spent in physical activity related to farm production as a measure of actual contribution. People who were

Table 7. Contribution of Men, Women, and Children as Percentage of Total Time in On-Farm Agricultural Operations.

Operation	Household				Local Hired				Migrant	Total
	Male	Female	Male Child	Female Child	Male	Female	Male Child	Female Child		
Plowing	28.6	0	0	0	61.3	0	0	0	10.1	100
Planting	22.0	5.4	10.4	5.3	56.7	0	0	0	0	100
Fertilizing	17.0	3.6	7.6	3.0	68.5	0	0	0	0	100
Weeding	3.1	.4	1.6	1.2	20.2	32.2	17	18.7	5.1	100
Herbicide	46.9	0	0	0	53.0	0	0	0	0	100
Rodent	84.6	0	0	0	15.3	0	0	0	0	100
Mechanical harvesting	26.8	.3	0	0	56.0	0	0	0	16.7	100
Manual harvesting	5.1	1.7	1.6	1.5	13.8	11.1	7.6	6.7	50.4	100
Threshing	21.2	2.0	5.6	3.0	22.7	1.2	1.8	1.8	40.3	100
Winnowing & cleaning	3.1	.4	1.6	1.2	20.2	32.2	17.1	18.7	5.1	100
Bagging	1.4	.2	0	0	39.9	4.3	4.0	4.3	45.5	100
Transport	17.9	.8	2.7	.4	23.8	5.0	4.0	4.2	40.9	100
Percentage	13.1	1.6	2.8	1.5	30.5	7.6	5.1	4.8	32.5	100

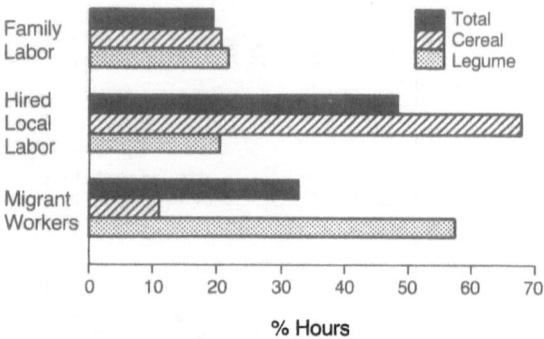

Figure 1. Contribution of Labor Categories as Percentage of Total Hours in Agricultural Activity.

not physically involved in any specific task (e.g., supervisors, farm managers) were not considered as contributors.

In Table 7, we show the contribution of men, women, and children to different agricultural operations. By examining labor input within the household, we see that household labor contributed 19.2% of the total labor input, much less than the contribution of hired labor (Figure 1). Females in the household provided only 3.2% of the total labor input. To maintain the prestige of the family, farmers do not allow their wives or daughters to work outside the household. In addition, some farmers were unwilling to report that their family members worked in the fields.

The contribution of hired labor was 80.7% of total labor input, with local hired labor more important than migrant workers. Hired local men, whose labor input was 30.5%, worked in skilled labor operations such as plowing, planting, and mechanical harvesting, while hired local women, whose labor input was 7.6%, were usually employed in manual operations such as weeding, manual harvesting, winnowing, and cleaning.

Migrant workers contributed 32.5% of the total labor input. In Table 8, we show the number of workers engaged in each farm operation. In terms of percentages, migrant workers were most needed for manual harvesting, transportation, threshing, and hand weeding.

The introduction of mechanical harvesting displaced women and children from agricultural operations. At the same time, the introduction of fertilizer slightly increased the labor input of household women as well as the need for hired labor. However, the adoption of herbicide, while increasing the need for hired labor, had a negative impact on the labor input of hired women and children.

Manual harvesting required the most labor (Table 9), as it amounted to 42% of the total labor hours spent on the farm.

Table 8. Contribution of Men, Women, and Children as Number of Workers in On-Farm Agricultural Operations.

Operation	Household				Local Hired				Migrant	Percent
	Male	Female	Male Child	Female Child	Male	Female	Male Child	Female Child		
Plowing	28	0	0	0	65	0	0	0	5	4.7
Planting	76	15	50	19	70	0	0	0	0	11.2
Fertilizing	38	3	21	0	46	0	0	0	0	5.2
Weeding	34	22	42	8	23	2	0	0	63	9.4
Herbicide	29	0	5	0	34	0	0	0	0	3.0
Rodent	20	0	2	0	0	0	0	0	0	1.0
Mechanical harvesting	11	3	9	0	80	0	0	0	11	5.9
Manual harvesting	19	13	13	7	45	31	8	6	343	23.6
Threshing	8	3	0	0	60	3	3	3	78	2.7
Winnowing & cleaning	6	1	2	3	6	11	5	6	8	2.3
Bagging	27	8	11	11	60	2	3	3	85	10.2
Transport	37	10	14	6	79	9	6	8	142	15.1

Table 9. Contribution of Men, Women, and Children as Number of Total Hours in On-Farm Agricultural Operations.

Operation	Household				Local Hired				Migrant	Percent
	Male	Female	Male Child	Female Child	Male	Female	Male Child	Female Child		
Plowing	1396	0	0	0	2984	0	0	0	496	7.7
Planting	1281	316	609	308	3294	0	0	0	0	9.1
Fertilizing	666	142	300	120	2671	0	0	0	0	6.1
Weeding	60	8	32	24	390	620	335	360	100	3.0
Herbicide	224	0	0	0	253	0	0	0	0	.7
Rodent	176	0	32	0	0	0	0	0	0	.3
Mechanical harvesting	935	12	0	0	1951	0	0	0	582	5.5
Manual harvesting	1364	466	448	424	3680	2978	2040	1800	12451	42.2
Threshing	315	30	84	45	336	18	27	27	597	2.3
Winnowing & cleaning	60	8	32	24	390	620	330	360	100	3.0
Bagging	35	6	0	0	954	104	96	104	1087	3.7
Transport	1799	82	272	44	2396	504	408	424	4112	15.8
Total	8311	1070	1809	989	19299	4844	3231	3075	20525	100.0

Table 10. Contribution of Men, Women, and Children as Number of Total Hours in Cereal Production.

Operation	Household				Local Hired				Migrant	Percent
	Male	Female	Male Child	Female Child	Male	Female	Male Child	Female Child		
Plowing	888	0	0	0	2070	0	0	0	327	10.7
Planting	841	160	300	80	2643	0	0	0	0	13.1
Fertilizing	422	22	172	0	2614	0	0	0	0	10.5
Weeding	458	400	584	0	400	0	0	0	0	6.0
Herbicide	168	0	0	0	251	0	0	0	0	1.3
Rodent	60	0	16	0	0	0	0	0	0	.2
Mechanical harvesting	845	12	28	0	1951	0	0	0	566	11.0
Manual harvesting	60	90	0	0	720	1320	2040	1800	0	19.6
Threshing	4	6	0	0	64	104	96	104	0	1.2
Winnowing & cleaning	0	0	0	0	390	540	330	360	0	5.2
Bagging	135	15	63	12	237	18	27	27	408	3.0
Transport	516	12	192	0	1316	504	408	424	2054	17.6
Total	4397	711	1355	92	12656	2486	2901	2715	3355	100.0

The Contribution of Labor in Cereal Production

In Table 10, we show the contribution of men, women, and children for differen activities in the production of cereal crops. We need to consider in greate detail specific agricultural operations and the contributions of differen categories of workers.

Farmers who own tractors prepare their own fields and sometimes do s(for relatives. Seven farmers in the sample owned tractors, while 86% of th(sample hired tractor operators and their machines.

Plowing usually occurred between March and May. Fifty percent of th farmers in the sample finished plowing in April. About 60% of the farmer used moldboard plows for this operation, while 20% used disc plows. Plantin; operations occurred between October and December. Most of the farmer (84%) planted wheat after the first rainfall. Eight percent of the farmers use(drills for planting, while 10% used chisel plows, 68% used disc harrows, an(16% used traditional plows pulled by animals to cover the seed. This operatioı was a man's task. Household and family members contributed about 27% of the total hours spent in planting, local hired labor 36%, and migrant worker about 10%.

Broadcasting was the most common method of planting. According to th survey, 74% of the farmers used the Hourani variety of seed, 16% used th F8 variety, and only 6% used the local variety. Thirty-eight percent of th farmers used stock seed from the previous year and 58% purchased seed fron cooperatives. The amount of seed per hectare ranged from 50 to 140 kilograms with the average around 100 kilograms.

Ninety percent of the farmers in the sample used either phosphate, nitrogen or mixed fertilizer. Forty-six percent of them used mixed fertilizer (an averag(of 100 kilograms per hectare), 22% used nitrogen (an average of 80 kilogram per hectare), 7% used nitrogen (an average of 60 kilograms per hectare), an(8% used nitrogen and phosphate separately (an average of 50 kilograms pe hectare for each). Application of phosphate or mixed fertilizer and plantin; occurred at the same time.

About 24% of the total labor time in cereal cultivation was spent in plantin; and fertilization (Figure 2). About 41% of these hours were contributed b local hired labor (Table 10), mostly male.

Weeding cereal crops by hand required 6% of the total hours spent in al activities. Family labor contributed about 80% of the labor time. Male childreı were the most frequently employed, because weeding does not require skille(labor.

The application of herbicide, a man's job, occurred during March and April Family labor contributed about 40% of the hours needed to perform thi operation, with local hired labor contributing about 60%.

Males were also responsible for the control of rodents. Men or boys fron the household walked through the field, covering every rat hole they saw and injected poison into any holes that reappeared the next day. This operatioı

AGRICULTURAL
OPERATION

Figure 2. Percentage Distribution of Total Hours for Agricultural Operations in Cereal and Legume Production.

occurred two or three times every 10–15 days.

Farmers with combines harvested their own fields and hired out their services to other farmers. Six percent of the farmers in the sample owned combines.

The cereal harvest began in the first week of June and continued until early July. Family labor contributed about 25% of the total labor hours for mechanical harvesting.

Farmers who owned livestock harvested by hand rather than machine in order to benefit from the straw. Hired child labor contributed about 60% of the total labor hours for this operation (Table 11). When a farmer decides to harvest the crop by hand, he usually hires the families of landless farmers and pays them in kind. The same families are responsible for threshing, cleaning, and winnowing, or the farmer may hire migrant workers to do the same jobs.

The operations after harvest (bagging, transportation) consumed about 21% of the total labor hours spent in cereal production. Farmers viewed bagging as men's work because of the physical strength needed to throw the bags onto trucks. About 40% of the workers in cereal production were engaged in this operation, half of them migrant workers.

Table 11. Contribution of Men, Women, and Children as Percentage of Total Time in Cereal Production.

Operation	Household				Local Hired				Migrant	Total
	Male	Female	Male Child	Female Child	Male	Female	Male Child	Female Child		
Plowing	27.0	0	0	0	63.0	0	0	0	9.9	100
Planting	20.8	3.9	7.4	1.9	65.6	0	0	0	0	100
Fertilizing	13.0	.6	5.3	0	80.9	0	0	0	0	100
Weeding	24.8	21.7	31.7	0	21.7	0	0	0	0	100
Herbicide	40.0	0	0	0	59.9	0	0	0	0	100
Rodent	78.9	0	21.0	0	0	0	0	0	0	100
Mechanical harvesting	24.8	.3	.8	0	57.3	0	0	0	16.7	100
Manual harvesting	.9	1.4	0	0	11.9	21.8	33.8	29.8	0	100
Threshing	1.0	1.5	0	0	16.9	27.5	25.3	27.5	0	100
Winnowing & cleaning	0	0	0	0	24.0	33.3	20.3	22.2	0	100
Bagging	14.3	1.5	6.6	1.2	25.1	1.9	2.8	2.8	43.3	100
Transport	9.5	.2	3.5	0	24.5	9.2	7.5	7.8	37.8	100
Percentage	14.3	2.3	4.4	.2	41.2	8.1	9.4	8.8	10.9	100

Table 12. Contribution of Men, Women, and Children as Number of Total Hours in Legume Production.

Operation	Household				Local Hired				Migrant	Total
	Male	Female	Male Child	Female Child	Male	Female	Male Child	Female Child		
Plowing	480	0	0	0	570	0	0	0	60	3.3
Planting	276	16	145	8	363	0	0	0	0	2.4
Fertilizing	64	0	8	0	57	0	0	0	0	.3
Weeding	802	564	948	228	1184	36	0	0	2084	17.7
Herbicide	10	0	0	0	2	0	0	0	0	.03
Rodent	56	0	16	0	0	0	0	0	0	.2
Mechanical harvesting	50	0	0	0	0	0	0	0	16	.2
Manual harvesting	787	376	448	424	2360	48	0	0	13211	53.7
Threshing	31	0	0	0	890	0	0	0	1087	6.1
Winnowing & cleaning	60	8	32	24	0	80	0	0	100	.9
Bagging	80	15	21	33	99	0	0	0	189	1.3
Transport	1091	70	80	44	1080	0	0	0	2058	13.4
Total	3787	1049	1698	761	6605	164	0	0	18805	100.0

The Contribution of Labor in Legume Production

In Table 12, we show the contributions of men, women, and children fo various operations in the production of legumes. Legume crops in the distric include lentils (Zone 2 only), chickpeas (Zone 1 only), and vetch. In our sample 54% of the farmers planted lentils and 6% of them grew chickpeas, but non planted vetch.

In preparing the fields for legumes, most farmers plowed once with a disc harrow during December. However, 16% of the farmers planting legume plowed twice, using a moldboard plow first, followed by a disc harrow. Fiv farmers in the sample owned tractors and carried out this operation b themselves.

Farmers planted lentils between late November and the first of January About 76% of them used their own stocks, while others bought their see in the market. Only one farmer in the sample used a drill for planting lentils He reported that the small size of the area planted in lentils and cereals in addition to the rental payment to the cooperative, did not encourage hin to use the drill. Farmers applied mixed and phosphate fertilizer at the sam time as they planted the seeds; this operation was similar to that for cerea crops.

About 23% of the farmers applied fertilizer to their legumes. Other farmer reported that fertilization has no impact on yield. Only one farmer reporte that he used herbicide for legume crops.

About 44% of the farmers used hired labor to perform these operations Other farmers were dependent on themselves and their family members Children helped farmers by carrying seed into the fields.

Weeding legume crops by hand was an activity shared by family members local hired laborers, and migrant workers, with the latter contributing mos of the labor hours (Table 12).

Two farmers in the sample harvested legumes by machine, one using combine for chickpeas, the other a self-propelled mower. Most legumes wer harvested by hand, traditionally a difficult and time-consuming task. A littl over half of the labor time for legume production was concentrated in manua harvesting (Table 13). Farmers harvested lentils during the second half o May. They reported that, in order to prevent losses, they were forced to hir labor to harvest the lentils as soon as possible. In fact, migrant worker contributed 74.8% of the total labor hours spent in this operation. Farmer reported that they didn't face any problem in hiring these workers, althoug some farmers had problems in managing the laborers in the fields. The numbe of workers in a field averaged about fifteen. Female participation was negligible except in the case of those farmers who were unwilling to pay cash for labo They used family members to harvest the crops.

Threshing was usually mechanized, and completed in many cases by hire labor (Table 13). Men did the winnowing and most of the bagging, whil cleaning seemed to be women's work. In regard to transportation, 95% o

Table 13. Contribution of Men, Women, and Children as Percentage of Total Time in Legume Production.

Operation	Household				Local Hired				Migrant	Total
	Male	Female	Male Child	Female Child	Male	Female	Male Child	Female Child		
Plowing	43.2	0	0	0	51.3	0	0	0	5.4	100
Planting	34.1	1.9	17.9	.9	44.9	0	0	0	0	100
Fertilizing	49.6	0	6.2	0	44.1	0	0	0	0	100
Weeding	13.7	9.6	16.2	3.9	20.2	.6	0	0	35.6	100
Herbicide	83.3	0	0	0	16.6	0	0	0	0	100
Rodent	77.7	0	22.2	0	0	0	0	0	0	100
Mechanical harvesting	75.7	0	0	0	0	0	0	0	0	100
Manual harvesting	4.5	2.1	2.5	2.4	13.3	.2	0	0	0	100
Threshing	1.5	0	0	0	44.3	0	0	0	54.1	100
Winnowing & cleaning	19.7	2.6	10.5	7.8	0	26.3	0	0	32.8	100
Bagging	18.3	3.4	4.8	7.5	22.6	0	0	0	43.2	100
Transport	24.6	1.5	1.8	.9	24.4	0	0	0	46.5	100
Percentage	11.5	3.1	5.1	2.3	20.0	.4	0	0	57.2	100

the total labor hours were contributed by men. The operations from threshing to transportation constituted about 22% of the total hours in legume production.

Farmers' Views About Mechanization

Any mechanization of legume crop cultivation will eliminate the need for most migrant workers and increase the incomes of farmers. Farmers were asked how they felt about using mechanical harvesters for legumes. All of the interviewed farmers were willing to adopt mechanization. They also added that mechanized harvesting of legumes costs less than harvesting by hand.

In a study at the University of Jordan dealing with the mechanized harvesting of lentils, Haddad (1986) found that the total grain loss by different methods of mechanical harvesting ranged from 129 to 342 kilograms per hectare (Table 14).

Farmers in the sample were asked what percentage of crop loss they would find acceptable if they used mechanical harvesting methods. Fifty percent of the farmers would adopt and continue to use mechanical harvesters with a 10% loss in total production, and 100% of them would accept a 5% loss in total production if they adopted mechanization.

Conclusions

The results of the study indicated that area planted in crops, family size, number of males, village site (Husun, Nu'aima, or Al-Shajara), and age and health of the farmer were statistically significant factors that affected labor input and the adoption of technology on sample farms in the rainfed farming areas of Irbid district.

We can conclude that agricultural production in these areas is dependent on hired labor. The following points summarize the labor contributions of men, women, and children:

1. Women contribute less than 20% of the total labor hours in agricultural production.

Table 14. Lentil Grain Loss Under Different Harvesting Methods.

Method	Loss (kg/ha)
Hand pulling	17
Rear–mounted tractor	129
Self–propelled axial	183
Grain combine with 2.4 m cutter bar	342

Source: Haddad (1986).

2. The contribution of family labor is also less than 20% of the total labor hours.
3. The average labor input of women from the household is 8% less than the total labor hours contributed by family labor, and 2% less than the total labor hours in agricultural production.
4. In most cases, locally hired men are the workers who operate the machines; they contribute about 50% of the total labor hours in agricultural production.
5. The input of local hired labor is greater in cereals than in legumes, but the input of migrant workers is greater in legumes than in cereals.
6. Migrant workers are mainly engaged in operations that call for unskilled labor and strenuous work.

References

Arabiat, S., Nygaard, D., and Somel, K. 1982. Issues of improving wheat production in Jordan. Amman: University of Jordan.

Arabiat, S. and Snober, B. 1984. Mechanization of agriculture and socioeconomic development in Jordan. Dirasat 12(7): 159–197.

Barham, N. 1986. (The climatic marginal limit for wheat production in Jordan.) In Arabic. Dirasat 13(8): 61–72.

Begum, S. 1985. Women and technology: Rice processing in Bangladesh. Pages 221–241 in Proceedings of a Conference on Women in Rice Farming Systems. Los Banos, Laguna (Philippines): International Rice Research Institute.

DOS (Department of Statistics). 1982. Agricultural sample survey, 1974–1982. Page 94 in Agricultural Statistics Yearbook and Agricultural Sample Survey, Amman.

DOS (Department of Statistics). 1983. Statistical Yearbook, No. 33, Amman.

DOS (Department of Statistics). 1986. Statistical Yearbook, No. 37, Amman.

Duwayri, M. 1985. Farming systems in rainfed areas. Pages 127–157 in The Agricultural Sector of Jordan. Amman: Abdel Hameed Shoman Foundation.

Haddad, N. 1986. Food Legumes Project in Collaboration with IBRC. Annual Report No. 5. Amman: University of Jordan.

HKJ (Hashemite Kingdom of Jordan). 1980. (The Five-Year Development Plan, 1976–1980, Agriculture). In Arabic. Amman.

Hourani, H. 1985. The supply of agricultural labor. Pages 69–87 in The Agricultural Sector of Jordan. Amman: Abdel Hameed Shoman Foundation.

Khasawneh, S. 1986. (The Jordanian labor force situation.) In Arabic. in Labor 36: 10–27.

Lipton, M. 1977. Why Poor People Stay Poor: Urban Bias in World Development. Cambridge: Harvard University Press.

Pinstrup-Anderson, P. 1979. Modern agricultural technology and income distribution: The market price effect. European Review of Agricultural Economics 6(1): 17–46.

Rassam, A., and Tully, D. 1986. Aspects of agricultural labor in northwest Syria. Pages 133–141 in Farming Systems Program: Annual Report for 1986. Aleppo: International Center for Agricultural Research in the Dry Areas.

Rassam, A. 1984. The Syrian Farm Household: Women's Labor and the Impact of Technologies. MA thesis, University of Western Ontario, London, Canada.

Richards, A. and Martin, P. (eds.) 1983. Migration, Mechanization, and Agricultural Labor Markets in Egypt. Boulder, Colorado and Cairo: Westview Press and the American University in Cairo Press.

Ruttan, V. 1977. The green revolution: Seven generalizations. International Development Review 19(4): 16–23.

Sulkharomana, S. 1983. Domestic resource costs of agricultural mechanization in Thailand: A case study of small rice farms in Supanburi. Pages 16–69 *in* Consequences of Small-Farm Mechanization. Los Banos, Laguna (Philippines): International Rice Research Institute.

Labor Use on Farms in Dry-farming Areas of Konya Province, Turkey

AHMET ERKUS, TANER KIRAL, HASAN TATLIDIL and
F. FUSUN TATLIDIL

Introduction

In spite of rapid growth in the industrial sector, agriculture has played an important role in the national economy since the beginning of planned development in Turkey. In fact, according to recent statistical data, agriculture has created employment for more than half of the working population, contributed 19.5% of the GNP, and produced almost 75% of the raw materials processed by the country's agro-industry. Moreover, the foreign currency obtained from the export of agricultural products has contributed to industrial development while agricultural production has satisfied almost all the demand for food by a population increasing year by year.

The vital role of agriculture in the national economy will increase with more efficient use of existing agricultural resources, which requires both technological and economic measures. The development of optimum production plans for individual farms and the mobilization of agricultural production according to these plans are extremely important for the future of Turkish agriculture.

There are 3.6 million farms in Turkey, nearly 99% of which are family farms under fifty hectares. Farm holdings are highly scattered and fragmented, and farmers have insufficient capital. These factors have obviously inhibited rational and profitable production, and thus the income levels of these farms are quite low. On the other hand, these farms have not operated according to optimum farm production plans. As a consequence, small family farms have sometimes experienced an increase in unemployment. However, during peak work periods farms use hired labor in addition to available family labor. According to the findings of case studies conducted at the farm level in various regions of the country, even though 42% of the rural population was unemployed, small family farms were using hired labor, while the rate of casual labor use was 10.8% when several farming activities overlapped during the same season (DPER 1967; GDP 1973, 1974, 1977; Aras and Cakir 1975; Erkan 1973, 1978; Cakal 1973; Erkus 1974, 1979; Karacan 1975; Akin 1975; Acil and Inan 1977; Demirci 1978; Karalar 1981).

These studies focused on an economic analysis of farming, with little emphasis on labor availability and labor use. In addition, they were conducted in various

Dennis Tully (ed.), Labor, Employment and Agricultural Development in West Asia and North Africa, 31–53.
© 1990 ICARDA.

regions of the country and on different kinds of farming activities. Thus, the levels of rural unemployment are not precisely known either under current farming practices or potentially with the application of advanced farm technology. Because of inadequate information on the levels of unemployment, it is impossible to take valid and effective measures to develop appropriate policies for solving the employment problem in the farming sector.

Since the foundation of the republic, there has been a gradual adoption and diffusion of new farm technologies, and the use of modern farming inputs has increased in agriculture. Nevertheless, comparing this progress with the situation in developed countries, one can easily observe that Turkish agriculture has not reached a sufficient level in several areas of agricultural technology such as mechanization, fertilizers, sprays for plant protection, veterinary medicine, and livestock breeding. We can point to inadequate farm capital and traditional ways of farming, particularly in the dry-farming regions, as the main reasons for the low levels of farm technology. Thus, an efficient use of available labor becomes difficult and, along with limited off-farm employment opportunities, there is a high level of unemployment on Turkish farms.

The main purpose of our case study is to determine the amount of available labor, the level of labor use, and the level of new technology implementation for a random sample of farms in the dry-farming areas of Konya province. In addition, we will examine the economic assets of the farms and the annual financial results of farming activities. We will also develop plans for optimum farm production and, in harmony with these plans, suggest possible changes in the use of labor for the farms in our sample. We conclude with some proposals for decreasing unemployment and increasing farm profitability.

Review of Previous Research

Few empirical studies of Turkish agriculture have focused on labor availability and labor use on farms. Some studies have only looked at the working conditions of farm laborers. For example, Pinar (1966) found that agricultural laborers in the Aegean region have worse working conditions, lower wage levels, and less social security than Turkish industrial workers.

Aksoy (1969) investigated the situation of farm workers from the standpoint of agricultural law and compared them with farm workers in some European countries. He pointed out that the working conditions of farm laborers differed significantly from those of their colleagues in other economic sectors, and called for new legislation to improve working conditions, wages, and social security.

Yalcin (1980) described the working conditions and problems of seasonal farm labor in the cotton belt of Cukurova region. The seasonal farm workers coming from different parts of the country to the region constituted 70% of the total labor force in farming. Among these seasonal workers, 20% were men, 54% women, and 26% were children. This seasonal labor force included small farmers (42%), landless farm workers (5.6%), and unemployed urban residents (52.1%). Yalcin also found that the seasonal laborers worked 10 hours daily, and that 78.9% of them had nutrition problems.

Turkel (1981) indicated that the contribution of family labor to the total labor requirements of the agricultural sector was 69.9%. He also noted that farm labor contributes 13.4% of the total labor requirement in agriculture. He stressed that improvements are needed in areas such as working conditions, wage levels, and social security.

Akin (1981) examined potential and seasonal unemployment in agriculture. In 1980, the number of potentially unemployed people in agriculture was 700,000; the highest level of seasonal unemployment occurred during December and February. In order to create new employment in agriculture, he recommended intensive farming, investment in irrigation, subsidized agricultural inputs, reduced fallow areas, improved pastures, and better methods of livestock production.

Irmak (1981) claimed that the high rate of unemployment in agriculture has resulted in increased migration from rural to urban areas.

Relevant to the concerns of our case study, Erkus et al. (1987) studied labor availability and use in the Ankara region. They found that labor availability varied between 3.42 to 4.75 manpower units among sample farms. The working period of available labor outside the farm varied according to the size of the farms: 0.29 manpower units on farms of 0.1-10 hectares, 0.25 units on farms of 25.1-50 hectares, and 0.42 units on farms of more than 50.1 hectares. On the other hand, the amount of idle or unutilized labor was highest (54%) for the smallest farms, and lowest (0.3%) for farms of 25.1-50 hectares; the average was 39% for all of the farms.

In spite of unemployment, these farms used hired labor during the peak work seasons. The percentage of hired labor varied between 3.5% and 18.2% depending upon farm size, averaging 11% for all farm size groups.

The research findings also showed that farms using plans for optimum farm production can reduce the rate of unemployment from 38% to 15%.

Esengun (1987) studied the same subject in the central district of Tokat province. He calculated manpower units per farm for various farm size groups: 4.5 (0.1-5 hectares), 5 (5.1-10 hectares), and 5.9 (more than 10 hectares), with 4.7 as an average for all farms. He also determined the unemployment rate to be 69.5% for farms 0.1-5 hectares in size and 9% for farms larger than 10 hectares. These farms also used hired labor, whose rates varied between 2.5% and 2.9% during peak work seasons.

Kiral (1987) determined labor and other input requirements of farms with various technological levels of production in central Anatolia. He found unemployment rates of 60% on farms in hilly areas and 64% on farms in the plains.

Plans for optimum farm production should make use of agricultural innovations, namely new ideas, methods, practices, or techniques that provide the means of achieving sustained increases in farm productivity and income. It is the extension worker's job to encourage farmers to adopt innovations of proven value. Research in Turkey on the adoption and diffusion of innovations started in the 1970s. Research studies focused on information sources for the diffusion of farm innovations, and the socioeconomic characteristics and communication behaviors affecting the decision-making processes of farmers in regard to specific farm innovations.

Akdas (1973) identified the information sources for the diffusion of commercial fertilizers among cotton growers. He found that farmers in his sample relied on other farmers within the same village (93.2%), the extension service (63.6%), radio (74.7%), commercial firms that supply farm inputs (14%), farm cooperatives (22.8%), and personal experience (13.1%) as sources of information.

In regard to the diffusion and adoption of chemical weed-killers in wheat production, Talug (1975) divided the adoption process into two stages, awareness and decision-making. He found that the most effective information sources for increasing awareness were farmers of the same village (46.2%), the extension service (32%), other farm institutions (10.9%), farmers of other villages (7.7%), and radio and television (0.8%), while the most effective information sources at the decision-making stage were the extension service (41.2%) and opinion leaders (39.5%).

Tatlidil (1978) conducted research on the diffusion and adoption of village development cooperatives in Eregli district, Konya province. He found the most effective information sources to be radio and newspapers, leaders of cooperatives, and the extension service at the awareness stage, and the latter two at the decision-making stage.

Methodology

In order to determine the sample for our case study, we selected three of the eighteen counties of Konya province that were most representative in terms of climate, topography, and the extent of dry farming. Second, considering the same criteria and collaborating with local officials of the Provincial Agricultural Directorate, we selected three villages for each county, a total of nine villages. All farms in these villages engaged in dry farming were accepted as the main population of the case study (751 farms). Taking actual farm size as the variable, we obtained a sample size of sixty-one farms by using the stratified random sampling method (Yamane 1967).

The sample farms were stratified into three farm size groups: (1) 0.1–10 hectares, (2) 10.1–25 hectares, and (3) more than 25 hectares. We determined the number of sample farms for each group to be 17 (Group 1), 35 (Group 2), and 9 (Group 3).

During October 1987, the research project team interviewed farmers and administered questionnaires for them to complete. We collected economic and financial data for the production year 1986–87. In addition, we obtained information on the diffusion and adoption of several farm innovations in the region over the last forty years.

In the analysis of the collected data, we estimated the amount of labor available in each farm size group and their averages for all sixty-one farms. In this analysis, we calculated the total labor force for each farm in terms of manpower units. In calculating available family labor, we subtracted the number of days when family members were out of work because of illness, education, and military service, as well as the number of days not permitted for farm activities on account of unfavorable weather conditions or religious and national holidays in the research area.

Thus, we estimated the total available working days annually, multiplying this figure by the available manpower units and calculating the annual working capacity of the farms in terms of manpower units. The number of workdays required by the enterprises of the farms and by housework, hired labor days, and the number of days the farmer and his family members worked in off-farm activities were combined to obtain the total number of days worked; by comparing this figure to the available labor of the farm, we determined the rate of idle or unutilized labor. In addition, we calculated the labor capacity in terms of hours of man labor in different production periods of farming. In this calculation, we assumed the working day to be eight hours due to the conditions of the research area.

After determining the labor availability and labor use, we took the functions of farm capital (assets) as a basis to examine the economic structures (Acil and Demirci 1984) and the annual financial results of the farms.

We developed the optimum farm production plans for all sample farms in each farm size category, in regard to existing conditions and the adoption of advanced farm technology (innovations). While designing the linear programming models for these farm production plans, we used both the average

Table 1. Limiting Factors of Production for Sample Farms.

	Farm Size Groups (ha)			
	0.1–10	10.1–25	25.1+	Average
Total farmland (ha)	6.04	18.25	44.59	18.73
Cultivated rainfed (ha)	5.98	18.16	44.58	18.66
Land for fruit and vegetables (ha)	0.01	0.06	–	0.04
Vineyards (ha)	0.05	0.03	0.01	0.03
Labor hours during sample periods				
3 March–26 April	1214	1346	1418	1327
27 April–2 Sept.	2854	3161	3316	3113
3 July–2 Sept.	1326	1476	1539	1450
3 Sept.–3 Jan.	2793	3114	3253	3060
Barn capacity	14.6	23.6	51.1	25.2
Sheepfold capacity	45.7	64.8	131.1	69.3

Table 2. Technical and Economic Factors Limiting Production on Sample Farms.

Crops	Maximum Production Level (%)
Cereals	67
Wheat	50
Barley	25
Oats	25
Pulses	33
Chickpeas	25
Lentils	25
Melons	10
Fodder crops	33
Cow vetch	25
Sainfoin	25

values of the data collected from our sample and the findings of former studies related to the subject (Acil 1963; Ozemir 1966; Akyildiz 1967; Kansu 1973; Erkus 1976; Inan 1977; Demirci 1978; Erkus 1979).

In the construction of the farm models, we accepted the following limiting factors in production activities (Tables 1 and 2): farm size, amount of labor available in terms of hours of man labor, the number of work days for different production activities (soil cultivation, maintenance, harvesting, etc.), size of sheepfolds and cattle barns (in square meters), crop rotation from the aspect of soil fertility, and market conditions.

The Research Area

Konya province is located in central Anatolia, from 36.5 to 39.5 degrees north latitude and from 31.5 to 34.5 degrees east longitude. The province is a vast plain, with 35.9% of the surface area consisting of mountains and high plateaus (MRA 1968). In general, Konya is deficient in water resources, with only Lake Beysehir of any importance in irrigation (MAFRA, 1987b).

The typical continental climate dominates the greater part of the province, which is characterized by notable temperature differences between seasons, and between day and night. However, some areas experience the typical characteristics of a Mediterranean climate. The annual average temperature is 11.5° C, with the highest temperature 40° C and the lowest –28.2° C. The province gets most of its precipitation during winter and spring, averaging around 315 millimeters annually (MAFRA 1987b).

The soils of Konya are mostly clay, silty, or a mixture of the two. The upper layers of soil have poor organic matter content and low natural fertility (Akalan 1971). The arable land of the province, including first, second, third, and fourth class soils, is 36%, which is higher than the average for Turkey (GDP 1973).

Konya is linked to surrounding provinces by five highways and one railway. These routes are always suitable for transport throughout the year. There is also a highly developed network of road connections between the counties and villages of the province. In general, farmers have been marketing their products within the boundaries of the province. Well-developed transportation facilities have positively affected farmers in producing for the market.

According to the results of the most recent population census (1985), the population of the province was 1,769,050. The percentages of urban and rural population were 48% and 52% respectively. The number of men and women in the population are almost equal. The rate of annual population increase is 2.5%. The rate of increase in urban settlements (4.7%) is greater than that of rural areas (0.6%) due to migration from the latter to towns and cities in the province.

According to the census, the average literacy level is 80%, with the literacy level of men higher than that of women (89%). In addition, 68.8% of the active working population (12 years of age and over) is employed in agriculture, 8.5% in industry, and the rest in other sectors (SSI 1987).

Land Use and Land Tenure

The total area of Konya is 5,085,700 hectares, consisting of cultivated farmlands (59.2%), pastures and meadows (19.5%), forest (13.9%), and land not used in agriculture, namely, lakes and marshy places, mountains, and settlement areas (7.4%). The farmlands are used for the cultivation of cereals, pulses, or industrial crops (95.6%), vineyards (1.9%), orchards (1.8%), and vegetables (0.6%) (MAFRA 1987b).

The great majority of farms are privately owned. In addition, some farmers also cultivate land as tenants or sharecroppers. The findings of a general village-level survey indicated that tenancy existed in 27% and sharecropping in 43% of the villages in Konya (MRA 1983).

Farm Size

According to data from the Provincial Agricultural Directorate, there are 194,837 farms in the province, with nearly 2,968,975 hectares of cultivated farmland (Table 3).

More than half (51.4%) of the farms in the province are small farms under 10 hectares in size. However, the operators of these farms have been cultivating 18.2% of the total farmland within the boundaries of Konya.

In addition to the small sizes of these farms, their holdings are scattered and fragmented, This situation has negatively affected the economic efficiency of the small farms. As a consequence of farm mechanization, farmers who own small and scattered plots of land and possess inadequate capital have

Table 3. The Distribution of Farm Size in Konya Province.

Farm Size (ha)	Number of Farms	Percentage	Total Cultivated Land (ha)	Percentage
.1–5	42,104	21.61	105,500	3.55
5.1–10	58,120	29.83	435,000	14.65
10.1–20	53,310	27.36	829,500	27.94
20.1–50	36,254	18.61	1,038,600	34.98
50.1 +	5,049	2.59	560,375	18.88
Total	194,837	100.00	2,968,975	100.00

Source: Provincial Agricultural Directorate, Konya.

preferred to run their farms on a sharecropping basis with farmers who have access to tractors and other farm equipment. Thus, the traditional way of sharecropping has been changing year by year as more and more small farmers have chosen this option during the last three decades in the region.

Use of Agricultural Inputs

The level of input use by farmers has gradually increased, especially during the last twenty-five years. These inputs include machinery, fertilizer, herbicides and pesticides, and high quality seed. During the 1960s, the numbers of tractors and combines in the province were 4,554 and 756 respectively (MRA 1968), increasing to 39,494 and 1,499 respectively in 1986. During this period, fertilizer use increased from 115,000 to 486,241 tons (the average for 1983–86), and the use of high quality seed exceeded 12,000 tons in the same period (MAFRA 1987b). Crop production depends largely on mechanization but in hilly areas the wooden plows are still in use. In 1986, the number of wooden plows was 14,011.

Crop and Livestock Production

Dry farming occurs on 85.7% of the cultivated land in the province. On only 30.7% of this area is land left fallow as part of the sequence of crop rotation. The major crops grown in the dry-farming areas are wheat, barley, oats, rye, lentils, and chickpeas; in the irrigated areas farmers grow sugar beet, dry beans, sunflower, potatoes, and alfalfa. Of the cultivated area, 87% is devoted to cereals, 6.4% to pulses, 5.7% to industrial crops, and 0.8% to fodder crops (Table 4).

In 1986, Konya farmers produced 372,215 tons of vegetables from 19,242 hectares of land, 217,682 tons of fruit from 8,470,259 trees, and 134,959 tons of grapes from 56,225 hectares of vineyards (MAFRA 1987b).

Table 4. Area Sown, Production, and Crop Yield in Konya (1986).

Crops	Area Sown (ha)	Production (ton)		Yield (kg/ha)
Cereals				
Wheat	959,900		2,419,945	2,520
Barley	609,902		1,706,234	2,800
Rye	29,520		51,707	1,750
Oats	27,610		53,479	1,940
Maize	1,001		2,930	2,930
Pulses				
Chickpeas	70,940		84,708	1,190
Lentils	40,855		30,922	760
Dry beans	8,420		7,404	880
Industrial crops				
Sugar beet	51,804		2,224,296	42,930
Cumin	30,810		10,865	350
Potato	6,520		130,245	19,980
Sunflower	16,638		13,798	830
Opium	770	Caps	532	690
		Seed	597	
Fenugreek	630		832	1,320
Hemp	55	Fiber	23	270
		Seed	15	
Sesame	785		161	200
Tobacco	5		5	1,000
Fodder Crops				
Alfalfa	6,546	Green	173,395	26,480
		Dried	33,269	–
		Seed	285	–
Cow vetch	7,406	Green	300	
		Seed	11,198	1,510
Sainfoin	934	Green	6,010	6,430
		Seed	23	

Source: MAFRA (1987b).

In 1985, the number of livestock in Konya province was as follows: 308,215 cattle (27% of them new breeds), 2,893,634 sheep (5% of them Merino), 682,854 goats (30% of them angora goats), 2,496,166 poultry, and 80,766 hives (69% of the modern type). Animal products for the same year included 34,953 tons of meat, 276,579 tons of milk, 4,099 tons of wool, 269 tons of mohair, 1,014 tons of honey, and 235,346,925 eggs (MAFRA 1987a).

Table 5. Types of Land and its Distribution on Sample Farms (per Household).

Farm Size Groups (ha)	Cropland		Vegetables/ orchards		Vineyards		Total Land in Use	
	ha	%	ha	%	ha	%	ha	%
0.1–10	5.98	99.04	.008	0.13	.05	0.83	6.04	100
10.1–25	18.16	99.52	.059	0.32	.03	0.16	18.25	100
25+	44.58	99.98	–	–	.01	0.02	44.59	100
Average	18.66	99.64	.036	0.19	.03	0.17	18.73	100

Table 6. Land Ownership and Tenure on Sample Farms (per Household).

Farm Size Groups (ha)	Land Owned		Land Rented		Land Sharecropped		Total	
	ha	%	ha	%	ha	%	ha	%
0.1–10	5.92	98.04	–	–	.12	1.96	6.04	100
10.1–25	15.36	84.16	.6	3.29	2.29	12.55	18.25	100
25.1+	26.78	60.06	4.33	9.72	13.48	30.22	44.59	100
Average	14.41	76.94	.98	5.25	3.33	17.81	18.73	100

Land Ownership, Land Use, and Land Tenure for Sample Farms

According to the results of our study, the average amount of land unde
cultivation per farm is 6 hectares for farms in the first group (0.1–10 hectares)
18 hectares for farms in the second group (10.1–25 hectares), and 445 hectare
for farms in the third group (25 hectares and over) (Table 5). More tha:
99% of the land in all groups is cropland. The rest is used for vineyards
vegetable gardens, or orchards, whose products are consumed by the farr
family.

The larger the farm, the greater the amount of land rented or sharecroppe
(Table 6). This is because the annual costs of owning a tractor or a dra
animal are higher than a small farmer can afford. Thus, the small farme
enters into a sharecropping arrangement with a farmer who has a tracto
and the traditional tenure system is transformed into a tractor contractin
system, as in other parts of the country.

In sharecropping, the owner provides the land, seed, and half of the fertilize
while the sharecropper meets the other costs of production; the crop is share
equally between both parties.

Population and Education Levels for Sample Farms

Turning to the population of the sample farms (Table 7), we find that th
smaller farms have fewer people because lack of land compels some famil

Table 7. Population by Age and Sex on Sample Farms (per Household).

Age groups	Farm Size Groups (ha)		
	0.1–10	10.1–25	25.1+
0–6			
Male	.29	.32	.67
Female	.18	.63	.33
Total	.47	.95	1.00
%	7.85	13.51	14.31
7–14			
Male	.88	.80	.55
Female	.82	.91	.55
Total	1.70	1.71	1.10
%	28.38	24.33	15.74
15–49			
Male	1.65	1.74	2.11
Female	1.35	1.66	1.67
Total	3.00	3.40	3.78
%	50.08	48.37	54.08
50+			
Male	.47	.54	.44
Female	.35	.43	.67
Total	.82	.97	1.11
%	13.69	13.79	15.87
Total (all ages)			
Male	3.29	3.40	3.77
Female	2.70	3.63	3.22
Total	5.99	7.03	6.99

members to migrate into the cities. As an average, 50.89% of the population is male and 49.11% female, with the majority of the total population in the 15–49 age group and the 7–14 age group. Ninety-two percent of the population over six years of age is literate; most illiterate people are women and the elderly (Table 8).

Family Labor on the Sample Farms

We calculated the available family labor for the sample farms in terms of manpower units based on the following coefficients for men, women, and children (Acil and Demirci 1984): 0.50 for girls and boys ages 7–14, 0.75 for women ages 15–49, 1.00 for men in the same age group, 0.50 for women over 50, and 0.75 for men over 50.

Table 8. Education Levels (Number per Household).

	Farm Size Groups (ha)			
	0.1–10	10.1–25	25.1+	Average
Primary school				
graduates and pupils	4.82	5.17	4.67	5.00
% of educated	96.40	91.34	82.51	91.24
Middle school				
graduates and pupils	.06	.20	.22	.16
% of educated	1.20	3.53	3.89	2.92
High school students	–	.03	.22	.05
% of educated	–	.53	3.89	.91
Achieved basic literacy	.12	.26	.55	.27
% of educated	2.40	4.60	9.71	4.93
Total educated	5.00	5.66	5.66	5.48

The manpower equivalents of the time out of work because of permanent illness, education, and military service were subtracted from the total number of manpower units. We determined that the farms have 3.5–4.0 manpower units, with an average of 3.8 units.

Men, women, and children work in different kinds of activities on the farms. Adult males are generally responsible for seedbed preparation, sowing, planting, fertilizer application, harvesting, and feeding animals. Women generally work in the home, but are also involved in hoeing, harvesting, and milking, while the children herd animals.

Farm Assets

Farm assets were classified according to their functions (Acil and Demirci 1984). The value of total assets ranges from 8,699,761 lira (TRL) for the smallest farms to 42,880,110 TRL for the largest farms, with 20,922,659 TRL as the average (Table 9) (1 USD = 932.85 TRL during 1986–87). Fixed assets made up the largest part of the total assets for the farms in our sample.

Land forms the major part of fixed assets for all farms, followed by buildings, with land improvements and plantings only 0.9% of the total. Livestock forms the major part of working assets, followed by machinery and equipment. Livestock is more important than machinery for the smallest farms, while the percentages of machinery and livestock in the other size groups are more nearly equal.

Current assets are very low, from 2.3% of the total investment for the smallest farms to 2.8% for the largest, with an average of 2.5%. The inadequacy

Table 9. Assets of Sample Farms.

| | Farm Size Groups (ha) | | | | | | | | |
| | 0.1–10 | | 10.1–25 | | 25.1+ | | Average | |
	Value (1,000) TRL	% of Total Assets	Value (1,000) TRL	% of Total Assets	Value (1,000) TRL	% of Total Assets	Value (1,000) TRL	% of Total Assets
Fixed Assets								
Land	3,508	40.3	11,126	52.4	28,714	66.7	11,598	55.4
Land improvement	24	.3	34	.2	0	0.0	26	.1
Buildings	2,856	32.9	3,420	16.1	3,556	8.3	3,283	15.7
Plants	283	3.2	95	.5	220	.5	166	.8
Total	6,671	76.7	14,675	69.2	32,490	75.8	15,073	72.0
Working and current assets								
Livestock	1,393	16.0	2,975	14.0	4,359	10.2	2,739	13.1
Machinery and equipment	389	4.5	3,026	14.3	5,029	11.7	2,587	12.4
Crops, feed, seed	181	2.1	382	1.8	744	1.7	379	1.8
Cash on hand	66	.8	154	.7	257	.6	145	.7
Total	2,029	23.3	6,537	30.8	10,389	24.2	5,850	28.0
TOTAL ASSETS	8,700	100.0	21,213	100.0	42,879	100.0	20,923	100.0
Liabilities								
Mortgages and accounts payable	162	–	905	–	305	–	10	–
Value of land rented or sharecropped	136	–	132	–	886	–	244	–
TOTAL LIABILITIES	–	100.0	–	100.0	–	100.0	–	100.0
NET WORTH	8,401		20,175		41,688		20,069	

of working and current assets as a part of total assets has a negative effect on the financial results of the farms.

The value of assets per hectare of cultivated land decreases as the farm size increases. Thus, while the value of assets per hectare was 1,441,550 TRL for small farms, it dropped to 1,162,440 TRL for medium-sized farms and 961,680 TRL for large farms. The average of this value was 1,117,010 TRL.

Financial Results of the Sample Farms

The average gross value of production (GVP) per farm increases with farm size (Table 10). In the first and second size groups, the major part of GVP

Table 10. Gross Value of Production for Sample Farms (per Household).

| | Farm Size Groups (ha) | | | | | | | |
| | 0.1–10 | | 10.1–25 | | 25.1+ | | Average | |
	TRL (1,000)	%	TRL (1,000)	%	TRL (1,000)	%	TRL (1,000)	%
Crops								
Field crops								
Cereals	504	29.9	1,535	41.4	4,558	61.0	1,693	45.8
Pulses	13	.8	128	3.4	–	–	77	2.1
Industrial crops	–	–	71	1.9	–	–	41	1.1
Fruits and								
vegetables	27	1.6	7	.2	47	.6	19	.5
Vineyards	55	3.3	13	0.4	11	.2	24	.6
Subtotal	600	35.6	1,754	47.3	4,616	61.8	1,854	50.1
Livestock								
Cattle	109	6.5	138	3.7	279	3.7	151	4.0
Sheep	179	10.6	336	9.1	365	4.9	296	8.0
Goats	36	2.1	86	2.3	42	.5	66	1.8
Poultry	16	9.7	14	.4	20	.3	16	.4
Increase in								
inventories	747	44.2	1,378	37.2	2,147	28.8	1,316	35.6
Subtotal	1,087	64.4	1,953	52.7	2,853	38.2	1,845	49.8
Total	1,687	100.0	3,707	100.0	7,468	100.0	3,699	100.0

comes from livestock, but for the large farms it comes from crops (61.8%).

On the average, most of the crop production value comes from cereals. The production values of pulses, industrial crops, vegetables, fruits, and vineyards decline as farm size increases.

Although livestock production is important for all sample farms, the cultivation of fodder crops has not been sufficient to meet the needs of animals.

Turning to the total variable costs of crop and livestock enterprises on the sample farms (Table 11), we see that the average variable costs per farm increase with farm size, while the variable costs of crop production are larger than those of livestock production for all of the farms. Total variable costs per hectare of cultivated land decline as the size of farms increases. In fact, while they were 81,200 TRL for small farms, they declined to 67,760 TRL for medium-sized farms, and 59,400 TRL for large farms, with 65,880 TRL as the average.

We calculated the value of the gross margin by subtracting the total variable costs from the gross value of production for the sample farms (Table 12). The average value of gross margin per hectare of cultivated land is 198,396 TRL for small farms, 135,396 TRL for medium-sized farms, and 108,095 TRL for large farms, with 131,616 TRL as the average.

Table 11. Variable Costs for Sample Farms (per Household).

| | Farm Size Groups (ha) | | | | | | | |
| | 0.1–10 | | 10.1–25 | | 25.1+ | | Average | |
	TRL (1,000)	%	TRL (1,000)	%	TRL (1,000)	%	TRL (1,000)	%
Crops	246	50.2	715	57.8	1,726	65.2	731	59.2
Livestock	244	49.8	522	42.2	923	34.8	503	40.8
Total	490	100.0	1,237	100.0	2,649	100.0	1,234	100.0

Table 12. Gross Margin Values (TRL) for Sample Farms (per Household).

| | Farm Size Groups (ha) | | | |
	0.1–10	10.1–25	25.1+	Average
Gross value of production	1,687,348	3,707,358	7,468,326	3,699,334
Variable costs	490,026	1,236,518	2,648,496	1,234,033
Gross margin	1,197,322	2,470,840	4,819,830	2,465,301

We calculated the gross return of the farms by adding the off-farm agricultural income to the gross value of production. According to the results of our study, off-farm incomes averaged 176,765 TRL for small farms, 63,071 TRL for medium-sized farms, and 0 for large farms, with an overall mean of 85,451 TRL. Thus, the off-farm agricultural income declines as the size of farms increases, with no such income in the largest size group.

The average value of gross return per hectare of cultivated land is 308,884 TRL for the smallest farms; it declines as the farm size increases. The average is 202,060 TRL.

The fixed costs on the farms consist of depreciation, repairs, and maintenance costs of buildings, wages of permanent laborers, the value of labor furnished by the farm operator and members of his family, interest on loans, cash rent, and the landowner's share in sharecropping arrangements. (These last three kinds of fixed costs were not included in total farm expenses during the calculation of net return, but were taken into account when computing family farm earnings.) According to the results of the study, fixed expenses per farm range from 1,020,087 TRL for the smallest farms to 3,271,014 TRL for the largest, with 1,917,965 TRL as the average.

The major item of fixed costs is the value of family labor for small farms and depreciation for medium-sized and large farms. Fixed costs per hectare of cultivated land decline as the size of farms increases, from 169,028 TRL for the smallest farms to 73,359 TRL for the largest, with an average of 102,395 TRL.

Net return (on total farm assets) is the difference between gross return and total farm expenses. Net return has a positive value and ranges from 539,217 TRL for small farms to 2,499,328 TRL for large ones, with 990,590 TRL as the average. Net return refers to the net earnings of the total capital invested (Acil 1956; Aras 1956). The ratio of net return to assets (total capital invested) reaches its maximum with 6.2% for the small farms, but is smallest (3.8%) for medium-sized farms, with an average of 4.7%. The net return value per hectare varies between 44,836 TRL (for small farms) to 89,348 TRL (for large farms), with an average of 52,885 TRL.

Family farm earnings per farm are 875,125 TRL for small farms, 1,314,913 TRL for medium-sized farms, and 2,495,191 TRL for large farms, with 1,369,291 TRL as the average. The family farm earnings of the largest farms are sufficient to cover the living expenses of a family of five.

Labor Use Under Current Conditions of Farm Production

The amount of labor available on the sample farms is 3.51 manpower units for small farms, 3.87 for medium-sized farms, and 4.03 for large farms, with 3.81 as the average. Under existing conditions, there are 268 work days per year in Konya province. We computed the number of man work days available per farm to be 940 for small farms, 1,037 for medium-sized farms, and 1,080 for large farms, with 1,021 as the average. In addition, the total number of man labor hours actually expended on the farms (both family and hired labor)

Table 13. Labor Use on Sample Farms (per Household).

	Farm Size Groups (ha)			
	0.1–10	10.1–25	25.1+	Average
Labor available				
Manpower units	3.51	3.87	4.03	3.81
Man work days	940.68	1,037.16	1,080.04	1,021.08
Man days worked on farms				
Family labor	208.45	314.37	378.55	294.60
% of on–farm	90.78	91.47	79.60	88.83
Hired	21.18	29.31	97.00	37.03
% of on–farm	9.22	8.53	20.40	11.17
Total man days	229.63	343.68	475.55	331.63
Man days worked off–farm				
by family labor	197.27	144.71	256.39	176.39
% of available	21.18	13.95	23.74	17.27
Total number of man				
days worked	428.90	488.39	731.94	508.02
% of available	45.59	47.09	67.77	49.75
Idle labor (man days)	511.78	548.77	348.10	513.06
% of available	54.41	52.91	32.23	50.25

and the number of hours worked in off-farm jobs were divided by eight to obtain the number of man days actually worked. We calculated the idle labor rate in terms of man work days by subtracting the number of days actually worked from the man work days available on the farms.

There is some idle or unutilized labor in all three farm size groups, but is largest for the small farms (Table 13). Although all farms have some unemployment, they also use hired labor for harvesting cereal crops. In general, the ratio of hired labor to total labor used has been increasing as the rate of intensification and the size of farms increase.

Labor Use According to Plans of Optimum Production

We used linear programming models to determine the optimum farm production conditions under current technology for the farms in our sample.

Table 14. Current and Optimum Organization of Production for Sample Farms.

	Farm Size Groups (ha)							
	0.1–10		10.1–25		25.1+		Average	
	Cur-rent	Opti-mum	Cur-rent	Opti-mum	Cur-rent	Opti-mum	Cur-rent	Opti-mum
Area (ha)								
Wheat	2.35	1.39	6.89	5.85	19.16	14.66	7.44	6.02
Barley	1.02	0.90	3.66	2.40	9.73	6.70	3.82	2.50
Rye	–	–	0.24	0	1.39	0	0.34	0
Lentils	–	0.60	0.47	1.82	0	4.46	0.27	1.87
Chickpeas	0.12	0.30	0.23	0.91	0	2.23	0.16	0.93
Sunflower	–	0.30	0.21	0.91	0	2.23	0.12	0.93
Cumin	–	–	0.16	0	0	0	0.09	9.09
Fallow land	2.49	2.49	6.28	6.28	14.30	14.30	6.41	6.41
Vineyards	0.05	0.05	0.03	0.03	0.01	0.01	0.03	0.03
Fruits and vegetables	0.01	0.01	0.06	0.06	–	–	0.04	0.04
Animal numbers								
Milk cows (local)	.59	1.00	.94	1.00	1.22	1.00	.89	1.00
Milk cows (hybrid)	.06	1.13	.11	1.41	.44	4.02	.15	1.72
Sheep	16.12	26.43	37.91	37.20	52.44	82.41	33.98	41.17
Angora goats	3.23	4.00	6.06	6.00	4.44	5.00	5.03	5.00
Poultry	8.94	8.94	9.80	9.80	15.78	15.78	10.44	10.44
Gross margin (× 1,000 TRL)	1,197	1,594	2,471	2,971	4,820	6,913	2,465	3,171
Increase of gross margin (%)		33.12		20.25		43.42		28.64
Idle labor (%)	54	44	52	42	0.3	0	39	35

In these models, we used the input-output data, the percentage of fallow land under current conditions, and the existing capacities of both barns and sheepfolds without change.

In Table 14, we compare the results of our calculations with the currently existing organization of production on the farms in our sample. The gross margin values for the optimum organization of farm production are greater than those for current production patterns in all of the farm size groups.

Under conditions of optimum farm production, the increases in gross margin values would result from an increase in the areas for cultivating lentils, chickpeas, and sunflower (grown in rotation), and the full use of barn and sheepfold capacity.

If the plans for optimum farm production were applied, farm income would increase and idle labor on the farms decrease. In fact, as Table 14 indicates, the unemployment rate would decrease from 54% to 44% on small farms, and from 52% to 42% on medium-sized farms; on the large farms, there would be no idle labor at all (these farms would use hired labor in addition to family labor).

Labor Use Under Conditions of Advanced Technology

To determine the effect of advanced technology on labor and incomes, we assumed the following conditions that differed from the present situation in the research area: (1) sainfoin, a perennial herb (*Onobrychis viciaefolia*), would be included in the production program; (2) the rate of fallow land, which varied between 32% and 41% under existing conditions, would decrease to 30% in the dry-farming areas; (3) barn capacity would be fully used by crossbred cattle; (4) thorough seedbed preparation, and the efficient use of fertilizers, herbicides, and insecticides would increase the cereal yields about 15%. Under advanced technology, the amount of fallow land would decrease between 0.7 and 0.9 hectare per farm, and an additional cow could be raised on farms in every size group. Farm income under the plans for optimum farm production would increase by 5% and the 35.3% rate of idle labor would drop to 15% as the average.

The Level of Agricultural Input Use

The small farms use less fertilizer and farm chemicals than the average for all sample farms. On the other hand, they consume more seed than the general average (Table 15). These figures seem to suggest that the small family farms (less than ten hectares) still operate according to traditional ways of farming in the region. In addition, the levels of fertilizer and farm chemical use are higher and the level of seed use is lower for farms larger than twenty-five hectares.

Table 15. The Levels of Agricultural Input Use in Wheat Production.

	Farm Size Groups (ha)			
	0.1–10	10.1–25	25.1+	Average
Fertilizer (kg/ha)	162	174	232	190
Seed (kg/ha)	240	224	210	225
Chemicals (g/ha)	1,653	1,779	1,900	1,777

The low level of agricultural input use among farms is mainly due to the reduction of public subsidies in recent years for several farm inputs such as fertilizers, farm chemicals, and seeds.

We also determined the level of farm mechanization for the sample farms studied. We found one tractor (varying between 45 and 70 HP) per 101 hectares of cultivated land, one combine drill per 270 hectares, and one fertilizer spreader per 251 hectares in the region. These data are indicators of extensive farming and a low level of farm mechanization.

Farm Innovations

In order to learn about the diffusion of various farm innovations in the research area, we interviewed the village headmen of nine villages. According to these men, tractors, combines, combine drills, seed cleaning, seed treatment, and seed selector first entered the area in 1951. Vaccination and medicines to treat animal diseases appeared in 1965, fertilizers, insecticides, and weed killers were first put into practice in 1969, and artificial insemination and the use of antibiotics in livestock production started in 1979.

All of these farm innovations were introduced to farmers by public organizations, such as the state farm supply agency, the state feed industry, the agricultural extension service, Department of Plant Protection, and the veterinary services.

Information Sources Used by Farmers

Farmers rely on several sources of information for learning about and then deciding to use farm innovations. At the awareness stage, extension workers are the most frequent information sources, followed by radio and television, and local farmers (Table 16). At the decision-making stage of the innovation adoption process, extension workers are again the most frequent information sources, but local farmers are more important than radio and television (Table 17).

Table 16. Information Sources Most Used at the Awareness Stage.

Information Sources	Number of Farmers	Percentage
Extension workers	23	37.7
Radio and television	18	29.5
Local farmers	13	21.3
Commercial firms (farm input suppliers)	5	8.2
Farm magazines	2	3.3
Total	61	100.0

Table 17. Information Sources Most Used at the Decision–making Stage.

Information Sources	Number of Farmers	Percentage
Extension workers	30	49.2
Local farmers	22	36.0
Commercial firms (farm input suppliers)	5	8.2
Radio and television	2	3.3
Farm magazines	2	3.3
Total	61	100.0

The percentage of farmers mentioning mass-media channels such as radio television and magazines decreased from 32.8% at the awareness stage to 6.6% at the decision-making stage. On the other hand, the percentage of farmer mentioning another person (extension worker or local farmer) increased from 59% at the awareness stage to 85.2% at the decision-making stage.

The findings of our case study in regard to radio and television as a source of information for farmers suggest that the mass media have still not been used effectively to support agricultural extension work in the country. Ou findings concerning other farmers as one of the most frequently used information sources is in harmony with the conclusions of another empirical stud (Tatlidil 1984), which recommended the contact farmer approach be effectivel used in agricultural extension work in Turkey.

Conclusions

According to the findings of this study, there is unemployment in every farm size group, but it decreases as the farm size increases. Nevertheless, the farmer hire labor during the peak work seasons. But in the other periods, farm famil members seek off-farm employment. In order to reduce the rate of unem ployment on the farms in our sample, we make the following recommendations

1. With the implementation of the plans for optimum farm production developed in this study, the rate of unemployment would be reduced by 10% on the small farms, 10.3% on the medium-sized farms, and 32% on the large farms.
2. With the plans for optimum farm production using advanced farm technology, the rate of unemployment could be decreased an additional 15% in comparison with the farm plans developed according to currently existing conditions of production.
3. The rate of unemployment could also be reduced by improving the quality of the pastures and meadows, by enlarging the growing areas of fodder crops, and consequently improving animal husbandry.
4. Regional agricultural extension services should take responsibility for diffusing plans for optimum farm production and introducing new farming techniques to farmers. The use of contact farmers could be an effective tool in agricultural extension work. In addition, radio and television can usefully supplement regional extension efforts by informing rural communities about new farming technologies.

References

Acil, A.F. 1956. (The Profitability of Tobacco Farms in Samsun.) In Turkish. Ankara University, Faculty of Agriculture, Publication 105, Ankara.

Acil, A.F. 1963. (Hazelnut Farms and Their Importance for the Turkish Economy.) In Turkish. Ankara: Ankara University Press.

Acil, A.F., and Inan, I.H. 1977. (Milk Production Costs and the Factors Affecting Them on Alpu Farms.) In Turkish. Ankara University Yearbook of Agricultural Faculty, Ankara.

Acil, A.F., and Demirci, R. 1984. (Agricultural Economics.) In Turkish. Ankara University, Faculty of Agriculture, Publication 880, Ankara.

Akalan, I. 1971. (The Characteristics of Salty Soils in Konya.) In Turkish. Ankara University, Faculty of Agriculture, Publication 434, Ankara.

Akdas, M. 1973. (The Adoption of Chemical Fertilizers and Their Effects on Cotton Farming.) In Turkish. Doctoral thesis, Ankara University.

Akin, B. 1975. (Economic Analysis and Optimum Organizations of Farms in Igdir.) In Turkish. Ankara University Publication 373, Ankara.

Akin, C. 1981. (Labor and Employment in Agriculture.) In Turkish. Pages 93–109 in Second Turkish Economic Congress, Izmir.

Aksoy, S. 1969. (Agricultural Law.) In Turkish. Turkish Association of Agricultural Economics, Publication 1, Ankara.

Akyildiz, R. 1967. (Foodstuff in Turkey.) In Turkish. Ankara University, Faculty of Agriculture, Publication 285, Ankara.

Aras, A. 1956. (Land Ownership and Land Tenure in Southeastern Anatolia.) In Turkish. Ankara University, Faculty of Agriculture, Publication 100, Ankara.

Aras, A., and Cakir, C. 1975. (Economic Analysis of Farms in the Gediz Irrigation Project Area.) In Turkish. Ege University, Faculty of Agriculture, Publication 211, Izmir.

Atalay, M. 1971. (The Influence of Climate on Soil Formation in Konya). In Turkish. Proceedings of the Fifth Scientific Meeting, Soil Science Association of Turkey, Ankara.

Cakal, F. 1973. (Determination of Optimum Organization of Farms in Erzincan.) In Turkish. Ankara: Sevinc Matbaasi.

52

Demirci, R. 1978. (Determination of Optimum Organization and Self-sufficient Farm Size for Cereal Farms in Kirsehir Province.) In Turkish. Doctoral thesis, Ankara University.

DPER (Department of Planning and Economic Research, Ministry of Agriculture). 1967. (Research on the Socioeconomic Structure of Farms.) In Turkish. Ankara: Guzel Sanatlar Matbaasi.

Erkan, O. 1973. (Research on the Determination of Self-sufficient Farm Size in Adana.) In Turkish. MA thesis, Adana University.

Erkan, O. 1978. (Farm Planning in the Asagi Ceyhan Irrigation Project Area.) In Turkish. Doctoral thesis, Adana University.

Erkus, A. 1974. (Income and Nutrition on Keskin Farms.) In Turkish. Ankara University, Faculty of Agriculture, Publication 542, Ankara.

Erkus, A., 1976. (Farm Planning in Tavsanli District Using Linear Programming Methods.) In Turkish. Ankara: Latif Matbaasi.

Erkus, A. 1979. (Research on Planning for Farms Receiving Controlled Farm Credits from the Agricultural Bank.) Ankara University, Faculty of Agriculture, Publication 709, Ankara.

Erkus, A., Kiral, T., and Erakton, S. 1987. (Labor Use on the Farms of Ankara Province.) In Turkish. Ankara.

Esengun, K. 1987. (Labor Availability and Its Evaluation on Farms in Tokat Province.) In Turkish. MA thesis.

GDP (General Directorate of Planning, Research, and Coordination, Ministry of Agriculture). 1973. (The Optimum Organization for Farms in Cumra District, Konya Province.) In Turkish. Ankara: Sark Matbaasi.

GDP (General Directorate of Planning, Research, and Coordination, Ministry of Agriculture). 1974. (Optimum Crop Patterns for Opium-growing Farms in Afyon.) In Turkish. Ankara: Sark Matbaasi.

GDP (General Directorate of Planning, Research and Coordination, Ministry of Agriculture). 1977. (Research on Optimum Crop Patterns for Farms in Erzerum.) In Turkish. Ankara: Kirali Basimevi.

Irmak, Z. 1981. (Agricultural Labor and Employment.) In Turkish. Pages 111–123 in Second Turkish Economic Congress, Izmir.

Inan, I. 1977. (Determining Sufficient Farm Size and Optimum Organization by Linear Programming.) In Turkish. Doctoral thesis, Ankara University.

Kansu, S. 1973. (Nutrients and Animal Nutrition.) In Turkish. Ankara University Faculty of Agriculture, Publication 492, Ankara.

Karacan, A. 1975. (Economic Analysis of Farms, and the Role of Farm Credit and Cooperatives in Community Development in Elazig Province.) In Turkish. Ankara University Faculty of Agriculture, Publication 176, Ankara.

Karalar, C. 1981. (Economic Analysis and Annual Results of Tobacco Farms in Izmir.) In Turkish. Izmir.

Kiral, T. 1987. (Research on the Determination of Physical Production Input Requirements of Enterprises on Farms in Ankara.) In Turkish. Ankara University Faculty of Agriculture, Publication 1001, Ankara.

MAFRA (Ministry of Agriculture, Forestry, and Rural Affairs). 1987a. (Briefing File.) In Turkish. Provincial Agricultural Directorate of Konya.

MAFRA (Ministry of Agriculture, Forestry, and Rural Affairs). 1987b. (Konya Agriculture During the Last Four Years.) In Turkish. Provincial Agricultural Directorate of Konya.

MRA (Ministry of Rural Affairs). 1968. (General Village Inventory of Konya.) In Turkish. MRA Publication 105. Ankara.

MRA (Ministry of Rural Affairs). 1983. (General Village Inventory of Konya.) In Turkish. Ankara: State Statistical Institute.

Ozemir, Y. 1966. (Labor Requirements, Labor Use, and Calculation of Costs in Working with Farm Machinery and Equipment.) In Turkish. Technical University, Istanbul.

Pinar, C. 1966. (Problems of Agricultural Labor and Wages in the Aegean Region.) In Turkish. Izmir: Istiklal Matbaasi.

SSI (State Statistical Institute). 1987. (Statistical Yearbook.) In Turkish. Ankara.

Talug, C. 1975. (Research on the Diffusion and Adoption of Farm Technologies.) In Turkish. Doctoral thesis, Ankara University.

Tatlidil, H. 1978. (The Diffusion and Adoption of Village Development Cooperatives in Eregil District, Konya Province.) In Turkish. Doctoral thesis, Ankara University.

Tatlidil, H. 1984. (Research on the Contact Farmer Approach in Agricultural Extension Work.) In Turkish. Ankara University Faculty of Agriculture, Publication 893, Ankara.

Turkel, S. 1981. (Labor Problems in Agriculture and Some Proposals.) In Turkish. Pages 81–92 *in* Second Turkish Economic Congress, Izmir.

Yalcin, O. 1980. (Research on the Socioeconomic Problems of Seasonal Farm Laborers in the Cukurova Region.) In Turkish. Doctoral thesis, Ankara.

Yamane, T. 1967. Elementary Sample Theory. Englewood Cliffs, NJ: Prentice-Hall.

Social and Economic Aspects of Decision Making Related to Labor Utilization and Choice of Technology: A Case Study of a Turkish Village

HALUK KASNAKOGLU, HALIS AKDER, A. ARSLAN GURKAN,
NUKHET SIRMAN, NAZIM EKINCI and MEHMET ECEVIT

Introduction

The expansion of rainfed farming that occurred throughout this century in Anatolia reached its limits in the early 1960s. Irrigated agriculture, which grew at a slower rate, started to replace this system from the 1960s onward. In spite of the existing potential for irrigation, rainfed farming will continue to be dominant. The expansion in cultivated area has been accompanied by new inputs and cultivating techniques that, in turn, have affected labor demand. Fertilizers, improved seeds, and irrigation have generally increased labor demand, whereas mechanization has reduced it. In addition, the introduction of new crop varieties has caused a rescheduling of labor activities.

These developments seem to have resulted in the seasonal labor demand of coastal regions, where irrigated agriculture prevails, being met by the labor surplus of other predominantly rainfed regions. The coexistence and competition of irrigated and rainfed farming pose new problems for agricultural policy. For example, a single support price for the same crop grown in both rainfed and irrigated systems is not sufficient for achieving income policy goals. Moreover, the image of economic development has usually been associated with irrigation and other modern inputs. In other words, researchers have neglected rainfed farming and its special problems, considering it 'traditional' and likely to be abandoned during the modernization process.

It is, therefore, of great importance to deal with the problems and constraints of rainfed farming systems. Kinik, the village we have chosen for our case study, reflects the characteristics of Anatolian villages in a region dominated by rainfed agriculture.

The Village Setting

Kinik is located nine kilometers southwest of the Ankara-Eskisehir highway and thirteen kilometers southeast of Sivrihisar, a major administrative and market town. It has a population of 335 individuals living in about 60 households. The majority of the people consider themselves to be natives of the region, although some families trace their origin to other parts of the

Dennis Tully (ed.), Labor, Employment and Agricultural Development in West Asia and North Africa, 55–78.
© 1990 ICARDA.

56

former Ottoman Empire, e.g., Syria or Bulgaria.

Links with Sivrihisar are still very strong today. About half of the Kinik families own houses in town, which they generally use during the period when their children are attending secondary school. Nine households have migrated to Sivrihisar, but return to the village every summer to cultivate their land. All the governmental bodies and trading firms with which Kinik people deal are situated in Sivrihisar, where people sell their produce, make purchases, and send their children to school. Nevertheless, the village is also linked to another market town, Cifteler, about thirty-three kilometers southwest of the village.

Forty women, daughters of present-day Kinik residents, have married outside the village, and about forty men, sons of Kinik householders, have permanen jobs outside the village. According to one of the older inhabitants, more than 400 people emigrated from Kinik between 1960 and 1980. Many of these migrants hold jobs in the civil service; workers and independent traders are fewer in number.

Methodology

To obtain the data for this case study, we first conducted an intensive interview with the village head and a farmer who was one of the first settlers in the village. Then we conducted a household survey of all the families in the village as the second phase of the study.

We processed results of the household survey and selected a sample of fifteen households for intensive interviews. The research team members visited each of these households and spent about four or five hours with each completing questionnaires.

Five households were subjected to additional in-depth interviews, where researchers stayed with the families several nights and thus also had a chance to observe the activities of family members. The interviews included questions related to the activities of the households as well as to the village in general.

Table 1. Monthly Rainfall in Sivrihisar, 1981-85 (mm).

	1	2	3	4	5	6	7	8	9	10	11	12
1981	76.7	49.2	36.4	36.6	47.7	13.1	10.1	0.0	20.3	42.8	40.4	24.0
1982	59.6	21.5	11.2	18.2	70.3	45.9	25.3	0.0	21.4	0.0	7.6	21.0
1983	40.9	30.4	40.8	38.4	25.5	33.1	20.0	3.1	1.3	1.7	10.9	37.3
1984	29.9	26.6	67.6	26.0	26.1	37.0	25.0	24.0	0.1	0.2	35.3	17.8
1985	56.6	56.6	19.5	28.0	42.2	30.6	5.0	21.8	0.0	80.2	35.3	50.0
1986	51.2	40.7	9.3	16.3	31.1	32.7	10.7	7.2	31.6	4.4	13.2	62.2
1987	49.7	19.1	48.7	51.0	32.7	33.6	47.7	8.1	2.9	31.2	28.6	74.2
1988	10.7	36.3	50.1	66.0	–	–	–	–	–	–	–	–

Source: State Meteorological Office, Sivrihisar.

Finally, we discussed and evaluated the responses to the questionnaires and impressions from the interviews with our initial informant, the longtime village resident.

During their visits to the village, research team members were able to observe the farmers at work and recorded their activities, in addition to conducting time and budget studies. We have also used information provided in the notebooks that we gave to the five households subjected to in-depth interviews; these notebooks were a record of the activities of the household members for a period of one month.

Climate

Kinik shares the arid climate typical of the rest of the central Anatolian plateau. Monthly rainfall is subject to important annual variations (Table 1). The rain requirements become critical between September and November and from March to early June. The 1981–82 season seems to be a 'good' year from the point of view of quantity and incidence of rainfall. In contrast, the 1982–83 season is a 'bad' one. Rainfall affected the average crop yields of wheat and barley during those two seasons (Table 2).

Rainfall is not the only climatic feature that affects yields in rainfed agriculture, however. Central Anatolia in general and Eskisehir in particular are known for harsh winters that may extend into spring. Sudden temperature

Table 2. Average Productivity in Eskisehir (kg/ha).

	Wheat	Barley
1980	2330	2310
1981	2300	2650
1982	2520	2560
1983	1870	1520

Source: Ministry of Agriculture, Eskisehir.

Table 3. Distribution of Village Agricultural Land.

	Area (ha)	Percentage
Residents	1150	36
Owned by former residents, no longer in agriculture	250	8
Owned by farmers in other villages who purchased it	100	3
Village pasture for use by Kinik residents only	1600	50
Village-owned land rented only to residents of Kinik for cultivation	100	3
Total	3200	100

58

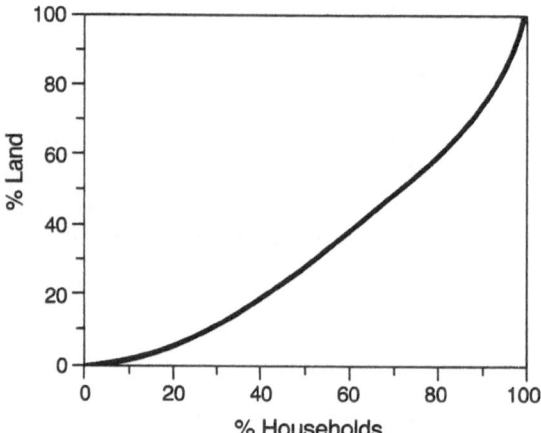

Figure 1. Distribution of Land Ownership in Kinik (Lorenz Curve).

drops in spring (sometimes to below zero centigrade) may quite effectivel reduce crop yields.

Resources (Land and Machinery)

Kinik's sixty households are surrounded by 3,200 hectares of agricultural lanc More than half of this land is village communal property (Table 3, Figur 1).

Around 1,600 hectares of land are available for cultivation. Farmers ca irrigate about 200 hectares of this land by pumping underground water fror a depth of 6 to 10 meters. Only about 1,500 hectares of land are availabl to Kinik farmers, as around 100 hectares are owned and cultivated by farmer from neighboring villages. However, Kinik farmers also own land purchase in neighboring villages. The amount of land (approximately 50 hectares) : small relative to what farmers can cultivate in Kinik.

Village residents own a total of forty-two tractors. Farmers estimate tha a tractor can easily cultivate 100 hectares a year. As the ratio of tractor to cultivated land in Kinik is approximately eighteen hectares per tracto this estimate seems to indicate over-mechanization as far as tractors a concerned.

There are four combines owned by Kinik residents. It is also possible t rent combines from outside the village.

In addition, there is a seed cleaner permanently based in the village, supplie by the Technical Agricultural Organization of the Ministry of Agricultur Farmers always use the seed cleaner to prepare their own wheat and barle seeds.

Cropping Patterns

In Eskisehir province, 87% of the total land area in any given year is used for wheat and barley production, most of it (80–85%) unirrigated. In Kinik, the picture is much the same. There are no alternative crops to wheat and barley that occupy land during all of the agricultural seasons. Sugar beet production, possible only on irrigated land that forms 12% of the total, is subject to strict supervision by the state-owned sugar company, the only purchaser of sugar beet. The company allocates land for sugar beet cultivation, which can occur on each plot only once every three years. Sugar beet is not an alternative to wheat and barley, because farmers can always use the land for cereal production after harvesting the sugar beet crop.

Chickpeas and cumin, which do not require irrigation, are usually grown as rotation crops on the same piece of land used for wheat and barley. Farmers know from experience that chickpeas grown on unirrigated land reduce soil fertility substantially for the wheat or barley cultivated immediately afterward. According to farmers, cumin is best grown on uncultivated land. In addition, the yield will fall substantially if cumin is grown in two successive years on the same plot of land. These considerations, coupled with low prices, caused Kinik farmers to abandon cumin in the 1960s. They started growing it again as a second crop after 1982 because of lower government prices for wheat and barley relative to the cost of agricultural inputs.

The main variety of wheat is the Russian soft wheat (Bezostia) originally supplied by the agricultural extension service. Some farmers are trying out a new variety, S7. The rest of the farmers are currently waiting to see the yield. This new wheat is especially suitable for irrigated land and yield expectations are quite high (8,000 kg/ha). In regard to barley, farmers grow a white variety suitable for beer production.

Their selection of either wheat or barley for cultivation is essentially governed by support prices of the Soil Products Office (TMO). However, farmers usually take into consideration what other farmers with adjacent land will be cultivating. This is because plot sizes are usually small (five hectares on the average) and it is very convenient for combines if adjacent plots have the same types of crop. Planting adjacent fields with the same crop also reduces the amount of potential contamination of a farmer's crop by other crops. For example, the price of wheat if contaminated by barley may be substantially reduced. Furthermore, the seed set aside during the harvest must be pure.

If farmers grow a rotation crop such as chickpeas or cumin, the next crop in the rotation sequence is usually barley. This is because the average yield of barley grown after cumin is higher than that of wheat.

Production Cycles of Main Crops

In Table 4 we show typical production cycles for wheat and barley. If a rotation crop is not grown, farmers harvest a crop once every two years on any given

Table 4. Timetable for Cultivating Wheat and Barley on Unirrigated Land.

| Mo/Year | Operation | Inputs | | |
		Equipment	Labor	Other
7/86	Harvesting of previous crop	Combine		
4,5/87	Cultivate fallow	Tractor, plow	Driver (.5 ha/hr)	15 l/ha fuel
6/87	Second plowing	Cultivator or harrow	Driver (1.5 ha/hr)	5 l/ha fuel
8,9/87	Preparation of seed	Seed cleaner, tractor	Operator, 1-2 helpers (6-7 tons/day)	
10,11/87	Seed bed preparation and seeding	Tractor-driven rake, combine drill	Driver (1.5 ha/hr) Driver, helper (1 ha/hr)	10 l/ha fuel Seed, fertilizer, chemicals (see Table 6)
3/88	Fertilization	Dispenser	Driver, helper (4-5 ha/hr)	150 kg/ha fertilizer
4/88	Weed defoliation	Sprayer	Driver, helper, (15-20 ha/day)	1 kg/ha herbicide
6/88 (Barley) 7/88 (Wheat)	Harvest and transport	Combine tractor, trailer	Driver (1-1.5 ha/hr)	20 l fuel

plot. Farmers could cultivate a given plot with wheat or barley every year but avoid doing so, as yields will fall dramatically. Currently, they prefer the cumin-barley rotation, which makes it possible to have two crops from the same plot in two years (Table 5).

The cultivation of wheat and barley is essentially mechanized. There is not much scope, in the opinion of farmers and local agronomists, for improving present technology except through increased irrigation and use of fertilizer. Irrigation on a large scale is only possible through government investment and farmers can do nothing except communicate their needs to the appropriate government agencies. On the other hand, they do take into account the advice of agronomists, who have recommended the use of ammonium sulfate instead of phosphorus-based fertilizers. Farmers whom we interviewed seemed willing to switch to the former in the coming seasons.

Another possible source of improvement would be the development of more suitable seeds. But farmers prefer to use their own (after processing in the seed cleaner) and complain about the high prices of government seeds, such as 150 lira (TRL) per kilogram for the same type of wheat they sell for 90 TRL per kilogram (1 USD = 1,700 TRL).

Given the level of mechanization in wheat and barley production, labor requirements are minimal, at most ten hours per hectare spread over two years. A farmer cultivating 15–20 hectares of wheat or barley needs 200 working hours or 20–30 working days. Although spread over two years, the timing

Table 5. Timetable for Cultivating Cumin-Barley on Unirrigated Land.

| Mo/Year | Operation | Inputs | | |
		Equipment	Labor	Other
7/86	Harvesting of previous crop			
11/86	Cultivate fallow	Tractor, plow	Driver (.5 ha/hr)	15 l/ha fuel
3/87	Seed bed preparation and seeding (cumin)	Tractor, rake, combine drill	Driver (1.5 ha/hr) Helper (1 ha/hr)	5 l/ha fuel 5 l/ha fuel 10 kg/ha seed 100 kg/ha fertilizer
4,5/87	Weeding	Sprayer	Driver, helper (15-20 ha/day)	1 kg herbicide
7,8/87	Harvest and transport	Tractor, trailer	2-3 persons 4-5 days	
9/87	Second plowing	Tractor, cultivator or harrow, rake	Driver (1.5 ha/hr)	5 l/ha fuel
10,11/87	Seed bed preparation and seeding (barley)	Tractor, combine drill	Driver (1.5 ha/hr) Helper (1 ha/hr)	5 l/ha fuel 200 kg/ha seed 150 kg/ha fertilizer 20 kg/ha pesticide
3/88	Fertilization	Dispenser	Driver, helper (4-5 ha/hr)	150 kg/ha fertilizer
4/88	Weed defoliation	Sprayer	Driver, helper (15-20 ha/day)	1 kg/ha herbicide
6,7/88	Harvest	Combine, tractor, trailer	20 l/ha fuel	

of operations may mean quite busy periods followed by idle ones. Cumin, on the other hand, requires around 180 hours per hectare over a six-month period. A farmer cultivating 4 hectares of cumin will need 720 hours or 80–90 working days. If the household meets all of the labor requirements, four workers will need 20–23 working days, most of which occur in July, the harvest season.

Cumin is harvested manually by all members of the household. If the plot is too big and/or there are not enough workers in the household, the farmer hires outside labor. The harvested crop is transported by tractor to a suitable threshing area near the house and spread out on the ground. The threshing operation is performed by driving the tractor, without equipment, over the crop, which at this stage consists of cumin and straw mixed together. The winnowing operation calls for a suitable (but not too strong) amount of wind. This operation may take several days, depending upon weather conditions. Members of the household are responsible for threshing and winnowing, although sometimes they may need help from other farmers or even hired labor if the harvested crop is large or if there are not enough available workers

The Economics of Crop Farming

Taking into account the unit input requirements for wheat, barley, and cumin (Table 6), we determined the costs of production for these three crops (Tables 7 and 8). Wheat and barley cost structures are quite similar except for the amount of seed applied.

In the classification of costs, we have used the categories of operating costs and imputed costs. Operating costs include costs of obtaining all inputs except labor in the case of wheat and barley, which do not require any labor from outside the household. All farmers in Kinik operate their own tractors and associated equipment. We have thus excluded labor costs from the operating costs category. In cumin production, we have also excluded all labor costs associated with operating tractors and equipment from the operating cost category. However, labor costs associated with harvesting are included in this category even if supplied from the household. Farmers, hard pressed for time sometimes hire outside labor to work side by side with their own in harvesting cumin. As we cannot separate these two sources of labor (one paid, the other unpaid) on a uniform basis, we have treated this category of costs as operating costs.

The treatment of imputed costs poses certain problems. For example, when a farmer rents a tractor for a specific operation, it comes with the required equipment and all operating costs are included in the rental price. In the preparation of these tables, we broke down this rental price into operating costs (items for which the farmer has to pay, such as fuel) and imputed cost associated with the use of his own equipment. We have estimated that 40% of the rental price can be attributed to operating costs and the remaining 60% to imputed costs for the use of equipment. Thus, for example, a rental price of 6,000 TRL per hectare in 1987 means that one would have to pay

Table 6. Input Requirements for Wheat, Barley and Cumin (per ha).

	Wheat and Barley	Cumin
Fuel	40 l (excludes harvest)	35 l
Seed	200 kg wheat or 250 kg barley	9 kg
Labor	10 hr (excludes harvest)	180 hr
Fertilizer	150 kg Diammonium Phosphate	100 kg 20/20
	150 kg Ammonium Nitrate	
Pesticide	16 kg cutworm treatment	
	16 kg seed treatment	
Herbicide	1 kg	1 kg
Combine	1.5 hr	

Table 7. Costs of Production: Wheat or Barley, 4 Hectares, 1987-88.

Operating Costs Input Type		Time	Amount or Rate	Total TRL
Fuel		4/87	60 l	12,000
Fuel		6/87	20 l	4,000
Fuel		10/87	40 l	8,000
Seed (wheat)		10/87	800 kg	80,000
Seed (barley)		10/87	1,000 kg	100,000
Fertilizer:				
Diammonium Phosphate (DAP) or		10/87	600 kg	72,000
Triple Super Phosphate (TSP)		10/87	600 kg	48,000
Chemicals (wheat)		10/87	3.2 kg	20,800
Chemicals (barley)		10/87	4 kg	26,000
Fertilizer		3/88	600 kg	67,800
Pesticide		4/88	4 kg	20,000
Fuel		3/88	20 l	7,800
Harvest		6/88		80,000
Transport[a]		7/88	100,000 kg	20,000
Subtotal (wheat)	(with DAP)			392,400
	(with TSP)			368,400
Subtotal (barley)	(with DAP)			417,600
	(with TSP)			393,600
Other costs (wheat and barley)				20,000
Total (wheat)	(with DAP)			412,400
	(with TSP)			388,400
Total (barley)	(with DAP)			437,600
	(with TSP)			413,600
Imputed Costs				
Rent (see text)		8/88	180,000 TRL/ha	720,000
Hired tractor[b]		4/87	6,000 TRL/ha	24,000
Hired tractor		6/87	2,500 TRL/ha	10,000
Hired tractor		10/87	3,000 TRL/ha	12,000
Hired tractor		3/88	3,000 TRL/ha	12,000
Own labor[c]		35 hrs	625 TRL/hr	21,875
40% interest on all costs excluding land rent & labor:				
Wheat	(with DAP)			140,160
	(with TSP)			130,560
Barley	(with DAP)			150,240
	(with TSP)			140,640
Total:				
Wheat	(with DAP)			940,035
	(with TSP)			930,435
Barley	(with DAP)			950,115
	(with TSP)			960,515
Grand Total (operating + imputed costs):				
Wheat	(with DAP)			1,352,435
	(with TSP)			1,318,835
Barley	(with DAP)			1,387,715
	(with TSP)			1,356,115

[a] Based on yield of 2500 kg/ha.
[b] Excluding fuel costs and driver's salary.
[c] Excluding five hours spent in transportation.

Table 8. Costs of Production: Cumin, 4 Hectares, 1987-88.

Operating Costs Input Type	Time	Amount or Rate	Total TRL
Fuel	11/86	60 lt	12,000
Fuel	3/87	40 lt	8,000
Seed	3/87	3.6 kg	3,600
Fertilizer	3/87	400 kg	40,000
Pesticide	4/87	4 kg	18,000
Fuel	4/87	20 lt	4,000
Weeding	4/87	30 persons	150,000
Harvest	7-8/87	40 persons	200,000
Preparation of crop	7-8/87	10-15 persons	60,000
Subtotal			495,600
Other costs			15,000
Total			510,600
Imputed Costs[b]			
Hired tractor[c]	11/86	6,000 TRL/ha	24,000
Hired tractor	3/87	3,000 TRL/ha	12,000
Hired tractor	4/87	2,500 TRL/ha	10,000
Own labor[d]	30 hrs	625 TRL/hr	18,750
40% interest on all costs excluding land rent & labor			52,640
Total			117,390
Grand Total (operating + imputed costs)			627,990

[a] Weeding, harvesting, and preparation of crop are usually performed by household labor, bı in some cases outside labor may be employed. Thus we include these items under operatin costs rather than imputed costs.
[b] Rent is not included among imputed costs, as rent payable in the summer of 1988 has alread been taken into account in the costs of wheat and barley. Farmers rent land for two yeaı (1986-88) and cumin is a second crop from the same plot of land.
[c] Excluding fuel costs and driver's salary.
[d] For operating tractor and associated equipment. Excludes labor for weeding, harvesting, an preparation of crop accounted for under operating costs.

10,000 TRL per hectare to rent a tractor to plow fallow land.

Interest rates are another source of difficulty. Commercial bank rates aı usually quite high, and the interest charged on agricultural credits (usuall for spring fertilizer and fuel) by the state-owned agricultural bank is subsidize at the rate of 40% per annum. The availability of credit is sometimes subjec to nonmarket criteria (e.g., the timing of parliamentary elections). In th unofficial credit market, interest rates are even higher (in one case, 10% pe month for a farmer to obtain fuel for spring farming operations). Th opportunity cost of money is thus subject to important variations. Unde these circumstances, we have selected a moderate 40% for our calculations, a rate which in our view must be treated as a lower limit.

Accounting for rent also poses problems. We have used the actual price to be paid for rented village land in the summer of 1988. This price is paid for unirrigated land taken over in the summer of 1986, to be returned in the summer of 1988. At most, farmers can grow two crops on this land, the most suitable being the cumin-barley sequence. The figure we have used, 180,000 TRL per hectare paid at the end of the rental period for the 1986–88 season, involves a slight overstatement depending on the rate of compounding (which may be high) and on the rental price paid in advance. From conversations with farmers on rents, we got the impression that the rate depends in part on the liquidity of assets available to the owner of the land and the prospective tenant.

In determining the profitability of crop operations for wheat (Table 9), we computed surpluses and profits on the assumption that the net price of wheat will be 140 TRL per kilogram after harvest. The operating surpluses are 63%, 70%, and 75% of the yield per hectare for the three yield estimates. The operating surplus forms the basis of the income received by the farmers. These high surpluses are instrumental in understanding the nature of the accumulation that the farmers have experienced over the years. Most farmers admit that they have accumulated most of their wealth during the 1970s when input prices were quite low relative to crop prices.

In regard to economic profit, what farmers consider as a 'normal' yield, 2500 kg/ha, seems to be the break-even level of yields. If farmers impute their own land, labor, and capital costs in the manner we have, the profitability

Table 9. Surpluses for Wheat, Barley, and Cumin, 1988 Prices, According to Yield (per ha).

	Wheat			Barley			Cumin		
Yield (kg)[a]	2000	2500	3000	2000	2500	3000	2000	2500	3000
Operating costs,									
TRL (1,000)	103	103	103	105[b]	105	105	128	128	128
Total costs, TRL (1,000)	338	338	338	329	329	329	157	157	157
Revenue TRL (1,000)[c]	280	350	420	260	325	390	400	450	500
Operating surplus,									
TRL (1,000)	177	247	317	155	220	285	272	322	372
Operating surplus, kg	1260	1760	2260	1190	1690	2190	270	320	370
Economic profit,									
TRL (1,000)	−58	12	82	69	-4	61	243	293	343
Economic profit, kg	−430	90	590	−534	−34	465	240	290	340

[a] Yields under 'bad', 'normal', and 'good' weather conditions.
[b] Excluding costs associated with redundant activities in the cumin-barley sequence as shown in Table 5.
[c] Based on net price of 140 TRL/kg for wheat (after tax of 4%), and on prices of 130 TRL/kg for barley and 1,000 TRL/kg for cumin.

of crop operations would become too critically dependent on expected yields
But then there is always the hope of high yields and, as a rule, farmers are
optimistic, being accustomed to 'trading with nature,' as they put it. It is
also conceivable that their main concern is with the expected operating surplus

Both the operating surpluses and economic profits of cumin farming are
high (Table 9), the latter because we have not included rent in the total costs
This, in part, reflects the farmer's view of rent as payment after the last crop
is harvested on a given plot before it is left for fallow. The economic surplus
of barley (Table 9) is more susceptible to yield on account of its low price
compared to wheat. However, by the very nature of our treatment we have
to consider the cumin-barley sequence in calculating surpluses (Table 10)
If a successful second crop is planted, both types of surpluses can be substantia
at the end of the rental period. We must note, however, that cumin is subjec
to unexpected and violent price fluctuations because, unlike wheat and barley
it is not subject to government price supports. Kinik farmers say that they
will be satisfied with 1,000 TRL per kilogram for cumin (1987 price) in 1988
despite an annual rate of inflation no less than 60–70%. The sort of sudder
price fall we have in mind occurred for lentils, a favorite second crop unti
1985, which Kinik farmers have now completely abandoned.

We can briefly describe sharecropping arrangements, using wheat as ar
example. The sharecropper uses the land for a period of two years. The
landowner provides one half of the seed and fertilizer; all other services and
inputs are supplied by the sharecropper. After harvest, the two parties equally
share the proceeds. Of course, there is no rent paid in this arrangement and
the return to land is implicit in the share of the landowner (Table 11).

Animal Husbandry

Raising sheep is a year-round enterprise for 56% of the village households
We cannot obtain input coefficients for this type of activity with the same
precision as for crop farming. The essential input is labor; with the exception
of the hired shepherd, all labor requirements are met from the household

Table 10. Operating Surplus and Economic Profit According to Yield, Cumin-Barley Rotation
1988 (per ha)[a].

Barley yield, kg	2000	2000	2000	2500	2500	2500	3000	3000	3000
Cumin yield, kg	400	450	500	400	450	500	400	450	500
Operating surplus, TRL (1,000)	536	606	676	601	671	741	666	736	806
Economic profit, TRL (1,000)	271	341	411	345	406	476	401	471	541

[a] We added 40% interest to the operating surplus and economic profit of cumin, then added
the resulting figures to the corresponding figures for barley from Table 9.

Table 11. Surplus and Profit to Sharecropper and Landowner per Hectare, Wheat.

Sharecropper			
Yield, kg[a]	2,000	2,500	3,000
Operating cost, TRL	84,100	84,100	84,100
Total costs, TRL	131,510	131,510	131,510
Revenue, TRL[b]	140,000	175,000	210,000
Operating surplus, TRL	55,900	90,900	125,900
Operating surplus, kg	400	650	900
Economic profit, TRL	8,490	43,490	78,490
Economic profit, kg	60	310	560
Landowner			
Yield, kg[a]	400	450	500
Operating cost, TRL[c]	19,000	19,000	19,000
Total costs, TRL	26,600	26,600	26,600
Revenue, TRL	140,000	175,000	210,000
Operating surplus, TRL	121,000	156,000	191,000
Operating surplus, kg	860	1,110	1,360
Economic profit, TRL	113,400	148,400	183,400
Economic profit, kg	810	1,060	1,310

[a] Yields under 'bad', 'normal', and 'good' weather conditions.
[b] In computing revenue, we assume the net price to be 140 TRL per kg.
[c] The landowner's operating costs include only half of the payments for seed and fertilizer.

Both men and women take part in feeding, watering, and cleaning barns. Women are responsible for milking and the production and marketing of dairy products. It takes two to three hours per day to water and feed the animals and to clean their barns. In the milking season, women of the household are busy four to five hours per day. Sheep raising thus fills the gaps in time that arise in mechanized crop farming.

Using the costs for raising a flock of 60 sheep as an example (Table 12),

Table 12. Revenues and Costs for Raising Sixty Sheep.

Revenues		Costs	
Item	TRL	Item	TRL
Cokelek[a]	120,000	Shepherd	120,000
Butter	30,000	Barley (feed)	960,000
Cheese[b]	420,000	Medicine	80,000
Wool	150,000		
Yogurt	300,000		
Lamb	1,200,000		
TOTAL	2,220,000		1,160,000

[a] Cokelek is a cheese product made from milk after butter is removed.
[b] Farm families consume about half of the total production of cheese and wool, and all of the yogurt.

if we exclude that part of the revenue imputed to self-consumption, we obtain a revenue of 1,635,000 TRL. This results in a cash surplus of 475,000 TRL. This amount is clearly far short of any compensation for the labor time expended on it, even if we allow for probable underestimation, considering the fact that the current daily wage rate is 5,000 TRL per day. The above-mentioned amount corresponds to 95 working days or around 855 working hours. In the milking season alone, two or three women work for four or five hours a day for 100 days on the average, roughly 1,125 working hours. What seems important to them is the cash they obtain in a period when no cash is forthcoming from crop operations of any kind. The main source of cash is the sale of lambs that takes place in April.

Cash Constraints on Farming

In crop farming, the proceeds of every harvest should cover all expenses incurred up to the next harvest. For example, the costs of preparing a crop for harvest in the summer of 1988 are paid by cash obtained in the summer of 1987. This practice may become burdensome during inflationary periods. For example, Kinik farmers bought phosphate fertilizer for 120 TRL per kilogram in October 1987; in June 1988, the price increased to 300 TRL per kilogram.

Consider a farmer who harvested ten hectares of wheat and ten hectares of barley in the summer of 1987 and who has the same area of wheat and barley to be harvested in the summer of 1988. Based on average yields, he obtained 25,000 kilograms each of wheat and barley in 1987. He had to keep 2,000 kilograms of wheat and 2,500 kilograms of barley as seed for the 1988 season. He had 23,000 kilograms of wheat and 22,500 kilograms of barley to sell. Prices of wheat and barley were 90 and 85 TRL per kilogram respectively in 1987. Out of 3,982,500 TRL of total revenue, the farmer had to pay for renting a combine and for transportation (268,250 TRL), leaving 3,714,250 TRL. To plant the same areas in the winter of 1987, he had to spend 1,175,000 TRL, excluding seeds and costs of transport and harvesting to be incurred in the summer of 1988. This would leave 2,539,250 TRL as subsistence for one year, corresponding to an income of about 211,000 TRL per month, for a typical farm family with five members. This sum compares favorably with urban wages (a university instructor gets about 200,000 TRL per month).

But other considerations can alter this optimistic picture. For example, debts contracted by the farmer that he had to repay by the summer of 1987 would reduce his basic cash fund. These debts may have been contracted during past production periods, or may have arisen out of a substantial investment in the purchase of land, agricultural equipment, or large scale expenses related to the repair of machinery or farm buildings, or even expenses related to the developmental cycle of the household – marriages, circumcision ceremonies, and the like. At least two farmers in Kinik had to repair their tractors in the summer of 1987 at a cost of 500,000 TRL or more. If the amount of debt reduces the farmer's basic cash fund below what is necessary for subsistence and the next agricultural cycle, his productive activities can redress the balance only under exceptional circumstances. The farmer may find himself locked into a cycle of cash constraint that leads to endemic indebtedness.

High interest rates and bad harvests may aggravate matters even further. These constraints help us to understand the farmer's need to plant a second crop, or the labor and time invested in animal husbandry.

The Social Organization of Farming: Introduction

Farming in Kinik is undertaken on a household basis. The household, under the management of its head, organizes production and consumption and is the main decision-making unit. All resources owned by the individuals who make up the household (land, agricultural equipment, animals, and available labor) are pooled under the management of the household head. Rather than ties of kinship or any other factor, it is this pooling of resources under the authority of the household head that defines the boundaries of a household.

Most household heads are male. Although households headed by females also exist (eight of the sixty households in Kinik are headed by women, all widows), these often rely on a grown-up son or other relatives to undertake agricultural tasks. Five Kinik widows have grown-up unmarried sons in their households and thus are able to enter into productive activities. The rest have to give their land to sharecroppers or individuals in other households who are often related to them through kinship ties (sons, sons-in-law, etc.).

Land Accumulation

Traditionally, households have obtained land through inheritance, sharecropping, and renting. As a result of land fragmentation, many farmers are left with land that is not considered sufficient for their needs, so that recourse to renting and sharecropping is widespread. It seems that in the previous generation, average landownership per household was fifty hectares. Today, the average is about twenty hectares per household, a figure that includes the land owned by all the individuals residing in the same household.

Nowadays, individuals can hope to inherit an average of 3.5 hectares. Our sample that includes 16% of the village shows that on average a household owns 17.2 hectares of land (see Table 13 for a distribution of resources among households in our sample). This includes land owned by all the individuals within the household, including the wife and live-in parents whose land is added to the household resource base. Households in our sample have on average inherited 7.2 hectares of land.

In order to establish viable farms themselves and leave something for their

Table 13. Distribution of Resources in Ten Sample Farms.

Land owned	Land rented	Land share-cropped	Total	No. of plots	Machinery	Sheep	Cattle	Workers
150	30	337	517	16	T FE A	70	7	4
50	167	107	324	10	T HE	25	6	2
150	110	135	395	9	T HE S	23	3	3
90	0	251	341	13	T HE	80	5	4
0	97	1203	1300	13	2 T, FE CH P	60	5	4
244	0	90	334	15	T FE CH	90	6	4
200	0	20	220	10	T+R	5	2	2
141	110	70	321	10	T FE	50	8	5
112	100	134	346	13	T FE	30	4	3
130	0	200	330	15	T FE S	35	4	2

Key: T: Tractor, FE: Full Equipment (trailer, plow, rake, flat plow, one-way disc plow, grain drill, and dispensers), HE: Half Equipment (usually plow, rake, one-way disc plow), CH: Combine Harvester, A: Automobile, P: Pickup, R: Trailer, S: Equipment shared with another household.

children, many farmers have to purchase land. Land purchasing is a phenomenon that began only ten years ago in Kinik.

The expenses involved in purchasing land, as well as inheritance practices that require the equal division of the patrimony among all children, regardless of gender, mean that Kinik villagers have to turn to other ways of ensuring their children adequate incomes. One important option is to educate children for jobs in the civil service or for semiskilled industrial jobs. Between 1960 and 1980, about 335 Kinik men left the village. These men often allow their share of the patrimony to be farmed by their uneducated brothers remaining in the village. Many of the sharecropping arrangements we observed in the village were of this kind. Educating women is much less frequent, because at marriage women join their husbands' households. Women who marry outside the village account for another large part of sharecropping contracts. As a result of these mechanisms, land purchasing accounts for a much smaller percentage of land ownership (4.7 hectares on the average) and half of the households in our sample have not purchased any land.

Division of Labor

Village farmers obtain the major part of their labor requirements from the members of the household. The core of a household is a man, his wife, and children. The man, as household head, manages agricultural activities and undertakes all the operations that are needed to complete a crop cycle. Management includes making all the decisions regarding the production of agricultural crops, decisions ranging from the purchase and maintenance of machinery to the sale of crops. Furthermore, men organize the sale of lambs in the spring.

The wife is in charge of all activities involved in processing animal products as well as managing their sale. Moreover, the wife is responsible for processing food, cleaning, preparing fuel from animal dung, and other tasks for the maintenance of the physical space within the boundary of the household. This space includes the sheep pen, a separate barn for cattle, the granary, outside kitchens, and other storage space. Husbands and wives cooperate in the cleaning and feeding of sheep and cattle. Women in general go to the fields only to harvest and to hoe labor-intensive crops such as sugar beet and cumin. Tasks that require the labor of more than one person in the cultivation of wheat and barley are often undertaken by the son of the household head and his father. If the labor of a young man is not available within the household, farmers try to enter into reciprocal labor arrangements with neighbors rather than use the labor of their wives or daughters. At times, a woman may have to supervise the planting of wheat or seed cleaning, which causes farmers considerable distress.

In the organization of work and division of labor, women's labor is continuous throughout the year. Their work is concentrated in the mornings and evenings when meals are prepared. Baking bread and washing clothes, undertaken once a week, represent the most time-consuming tasks. During a three-month period in the summer, making cheese and butter requires long and concentrated work on the part of women. For households that undertake such activities, harvesting and hoeing labor-intensive crops constitute the other period of intensive work that marks the women's seasonal work cycle. Men's work is more intermittent. For example, in the winter, men are mainly occupied in looking after animals, which does not take more than two hours a day.

Production and Consumption Decisions

The household is the main social unit that organizes consumption, and production decisions have to be taken in conjunction with the needs of household members. Food and clothing constitute a large part of household consumption. Kinik villagers purchase most of their daily consumption items (such as flour, vegetables, tea, soap, and clothing) in the Sivrihisar market. Meat, cheese, milk, and molasses, other important items of consumption, are produced by the household rather than bought. Including expenditures on health care, a Kinik household spends approximately 260,000 TRL on daily consumption items. As women are responsible for daily consumption, it is they who decide the amount to be spent on food and clothing.

Households also have to regulate social reproduction, the establishment of one or more viable households for the next generation. The costs of this generational reproduction have to be met through production. Buying more land in order to leave children sufficient land, improving the technological arsenal, educating sons, and organizing weddings in the socially acceptable fashion are therefore also among the perceived duties of the household head and his wife.

A wedding is an expensive affair. The parents of the bridegroom have to buy gold coins and bracelets for the bride, an expenditure that may exceed 500,000 TRL. This gold represents an important transfer of wealth from one generation to the next, a sum that the young couple often uses to build a house or to purchase a piece of agricultural equipment. Including the expenses of the wedding ceremony itself and the gifts exchanged between the affinal families, marriage of a son may cost about one million TRL. The marriage of a daughter can cost a similar amount of money because nowadays a bride's parents must provide an elaborate trousseau including household utensils, furniture, and even television sets. A woman will start preparing her daughter's trousseau and her son's gold from the moment they are born. The quantity and nature of the trousseau and the gifts are important reflections of the status of the family. Hence many of these reproductive expenses are under the direction of the woman rather than her husband, and it is she who makes many of the decisions.

Kinik households must meet the costs of production, consumption, and social reproduction by the income from agriculture and animal husbandry. We have noted that farmers require a total of 260,000 TRL to ensure daily subsistence needs. Assuming that farmers produce 2500 kilograms per hectare of wheat and that they sell it for 140 TRL per kilogram, this means that each household requires at least 1.05 hectares per person in order to meet these expenses after deducting operating costs. Therefore, a household of six will need about six hectares of cultivated land per year; allowing for fallow twelve hectares per household are necessary to meet daily consumption needs. Due to the difficulties of imputing a cash value for generational reproduction we have not attempted to calculate the amount of land needed to reproduce a typical household across generations. Many of these expenditures are met over a period of more than ten years, according to the needs of the particular family concerned. The farmers themselves say they need to cultivate at least twenty-five hectares per year.

In addition, the average Kinik farmer tries to raise at least forty to fifty sheep, as well as two or three milking cows, especially during the spring when cash funds have been exhausted. The proceeds for the sale of a calf or a few lambs can ensure the purchase of needed fuel, or can pay for repairs or even be used as the first installment in the purchase of a new piece of machinery. The safety valve role played by animal husbandry in the cycle of agricultural production for farmers whose main constraint lies in the shortage of cash is made even clearer by the fact that Kinik farmers with access to sufficient cash do not raise animals.

The Role of the Village in Decision Making

In addition to the household, the village is also an important unit in agricultural production. Through links established within the village, many farmers are able to obtain essential means of production that are not supplied by the household. Two areas of village cooperation are important: the allocation of land and the sharing of technology. We have seen that most farmers resort to renting and sharecropping arrangements in order to obtain the land they need. On the average, Kinik farmers sharecrop about 5.5 hectares and rent 6.1 hectares per household. The relations between landowners and their tenants/sharecroppers show that kinship and village ties play an important role in the allocation of land. Of the twenty-two Kinik households involved in sharecropping arrangements, twelve obtain land from kin (mother, father, sibling, and father's sister), eight from fellow villagers, and only two from residents of neighboring villages.

The village also acts as a significant source of land for farmers. The village as a corporate unit owns about 9.6 hectares. Each year this land is rented out to the highest bidder for a period of two years. Only farmers from the village are allowed to participate in the bidding. The income from renting village land is used to improve the village roads and water supply system. Six of the ten farmers in our sample rent land, 74% of which belongs to the village.

Machinery and Labor

Apart from land, households also exchange labor and agricultural equipment. Twenty households in Kinik do not own any agricultural machinery. Of the remaining forty, only seventeen own the full set of machines needed to undertake production (plow, trailer, seed planter, fertilizer dispenser, herbicide tanker, etc.). Another eight households share a full set with brothers living in other households, and fifteen households have only half the equipment they need. Borrowing equipment from neighbors is an informal affair as long as the farmer has his own tractor. In many cases, a person may even borrow a piece of equipment for the simple reason that he has lent his to someone else or because his is not ready for use. Farmers use many of these machines, which constitute a significant investment, for only seven days a year at most. Therefore, we can say that, taking the village as a whole, there is more machinery, including tractors, than is actually needed for the amount of land cultivated.

A more formal way of sharing equipment also involves labor exchange mechanisms. Planting, fertilizing, and applying herbicides require the cooperation of two people. Where the equipment needed to accomplish the task is owned jointly, farmers take turns in helping one another. In instances where the farmer does not own the machine, he obtains it in return for helping its owner. These forms of labor exchange involving the sharing of equipment can even be used for labor-intensive crops such as cumin and sugar beet.

Sharing labor and machinery occurs only at the village level. This type of cooperation reduces the amount of cash needed to complete the agricultural cycle in cases where the resources available within the household are not sufficient. Nevertheless, cash is not totally absent in relations between households within the village. Farmers without tractors can rent them from fellow villagers at the current market rate. A labor market also exists within the village; families in need of cash can send laborers to harvest cumin and sugar beet for other village households. Very few families, however, want to work on the land of nonvillagers; moreover, a household head rarely accepts such work, as he sees it as incompatible with the status of independent farmer. Villagers often have to resort to hiring outside labor for hoeing and harvesting sugar beet. They recruit from villages in Konya province at least one hundred kilometers away. These are people with experience in sugar beet cultivation. They ask for a lump sum of money, live on the fields in their own tents, and do not leave until they finish the job. In 1987, a husband-and-wife team from Konya took 60 days to harvest a 4.5-hectare sugar beet field for a piece rate of 120,000 TRL per hectare.

Other Examples of Village Cooperation

Village cooperation is also a significant aspect of animal husbandry. Households with sheep must hire a shepherd for about nine months of the year. These households often hire shepherds jointly, as one shepherd can look after at least 500 sheep and individual households do not own more than 60–90 sheep. In Kinik, three shepherds look after 1,600 sheep owned by 32 households. Households take turns in feeding and lodging the shepherd who is paid according to the number of sheep each household owns. The village hires another herder to graze cattle, but he is paid by individual families according to the number of animals they have.

Although households as units of production and consumption cooperate in production activities, there is little cooperation when it comes to marketing. Farmers have to deal independently with merchants, government officials, and banks. Many agricultural products are sold either to merchants or to government bodies, depending on the price offered and on the amount of indebtedness.

Farmers sell animals to buyers in Sivrihisar, or to itinerant sheep drovers who come to the village in spring and in autumn. Each week, women take milk, butter, and cheese to Sivrihisar by bus to sell to regular customers, either corner grocery shops or women in the town.

Information regarding prices, whether for animal or agricultural products, circulates widely in the village. Through such channels of information farmers obtain the necessary knowledge for making decisions. A novelty can only be adopted through the efforts of a farmer trusted by villagers, and demonstration of success is the most effective means of introducing new techniques. (Farmers are waiting to see the results of planting S7, a new variety of wheat, before using it themselves.) Farmers in need of cash are more prepared to try out new methods of cultivation than those with a comfortable income. Such farmers will try a new method or a new variety of seed on one of the many plots they cultivate. Large farmers involved in trading activities are also ready to take risks and can thus act as innovators.

Kinik households can, with the resources available to them and the mechanisms of cooperation established at village level, enter into productive activities in order to reproduce and enlarge their resources. Cooperation that exists between households does not entail the sharing of profits, although in some limited cases, such as sharecropping, risks can be shared. Farmers share assets such as machinery only to the extent that they find it mutually advantageous; once again, there is no sharing of profits.

Kinship ties facilitate mutual help mechanisms between households, but even in such cases people keep records of expenditure and expect debts to be paid. For example, in one case, a sixty-year-old man who had educated his three sons bought a tractor in 1982 with money they had loaned him. In return, he has been sending them an average of 100,000–150,000 TRL each per year. The kinship tie in this case meant that the sons could not easily refuse their father the money he needed and that the father could repay the debt in the manner that suited him best. Thus, although kinship often regulates the formation of households and determines who is going to live with whom, it does not necessarily organize cooperation and risk sharing between households. As with kinship, neighborhood and village ties serve to lubricate exchange and cooperation and do not create productive units larger than the individual household.

Land is a resource that cannot be shared by separate households to the same extent as machinery. Land represents the identity of a particular household and is the foundation for the establishment of an independent household. When a man gets married, for a number of years he remains under the tutelage of his father in whose name the land is still held. As the father gets older, the operation of the farm gradually shifts to the sons who have to cooperate in order to survive. Twenty-two of the sixty Kinik households sharecrop land. In seven cases, farmers sharecrop land that belongs to their fathers and an additional four have entered into sharecropping arrangements with their brothers who have left the village. Thus, many of the sharecropping arrangements are the direct result of the value placed on land and the rules of partible inheritance. Complete separation from the parental household means having to work as a laborer, an occupation that is avoided because it does not bring either sufficient income or status. Educating one or more sons is a means of preventing this eventuality. When the father dies, the land is divided among the heirs and brothers begin farming on their own.

Although markets in land are well developed, the sale of a piece of land

always involves other members of the family, because it is treated as the mos
important patrimony. Machinery, including tractors, is also part of th
patrimony included in inheritance calculations. But, as most holdings are we
below the capacity of any single tractor, it is possible to use a tractor jointl
and still keep separate accounts.

Mechanization and Its Consequences

Changes in technology account for many of the social arrangements observe
in the village. The introduction of tractors in the 1950s meant that a singl
household could farm more land than before. Increases in the cost c
production that have accompanied the introduction of tractors and the genera
rise in standard of living have, however, meant that farmers need even mor
land than before in order to ensure a sufficient patrimony for the nex
generation. They have readily adopted mechanization as a way of freein
male labor from agriculture, an option which has also served to counterbalanc
the negative effects of land fragmentation. By educating their sons, Kini
farmers have tried to optimize the distribution of their limited landholding:
Thus, practices relating to mechanization and investment in education ar
intimately related to the organization of land ownership. We should also not
that, except for combines, agricultural equipment represents a much smalle
investment compared to the purchase of land. Moreover, due to uncertaintie
in the availability of land to rent and to sharecrop, inheritance still remain
the one dependable form of access to land.

But the inability to depend on the labor of sons has also led Kinik farmer
to purchase equipment they will not use more than once a year, clearly
case of over-mechanization. Calculating their labor needs in terms of majc
agricultural tasks such as planting, harvesting, and plowing, farmers hav
sent their sons away at an age when they could have accomplished man
important but less visible tasks, such as filling the tractor-drawn fertilize
dispenser, or supervising seed distribution during planting. Neighbors perforr
such tasks on a mutual aid basis, a solution that may objectively result i
a better use of male labor but is, in the eyes of villagers, a constant sourc
of trouble as they must cooperate with equals rather than expect complianc
from dependents. The labor of sons is also necessary for looking after sheej
particularly when they return from pasture to be herded into the barns. On
shepherd for about ten households cannot do the job, which has been take
over by women.

Under present landownership patterns, the gender division of labor ha
also changed. Before mechanization, women's labor on the fields was les
frequent. With the introduction of cash crops, women's work load has increase
because the labor-intensive tasks associated with the cultivation of cumir
sugar beet, and chickpeas have become women's work. Men, who have alwa}
undertaken all the tasks involved in wheat and barley cultivation, have ha

their work load reduced significantly, while women have become the main source of labor for manual hoeing and harvesting. We must immediately point out that sending sons away for education has not affected a farmer's ability to cultivate labor-intensive crops, because daughters have remained at home to perform such tasks.

A further consequence of over-mechanization is that farmers now need to cultivate more land in order to pay for the equipment they have purchased as well as to employ productively all the horsepower available to them. It also seems that the cultivation of cumin and chickpeas and the raising of animals are undertaken in order to use labor that would otherwise be idle. As argued above, both these activities do not interfere with educating sons, but simply result in a transformation of the gender-based division of labor. Moreover, these activities produce cash when it is least available and most needed. Apart from optimizing labor use, the cultivation of summer crops such as cumin also helps increase land productivity because it allows the production of a second crop without affecting the basic wheat-barley cycle.

This analysis also suggests that farmers in Kinik are trying to maximize all the resources available to them without changing the basis of household formation, namely, independent access to land and other inputs. Farmers would readily adopt any change in technology that maintains this balance. We therefore suggest that finding ways of increasing the productivity of male labor (e.g., mechanization) is not as urgent as finding ways to increase the productivity of land. Irrigation, one way of doing so, could only be undertaken by the state. Improvements in seed and fertilizer are much more feasible. There could be new ways of utilizing male labor during the idle winter months. However, many projects that have attempted to do this through, for example, handicrafts production, have only resulted in overworking women instead.

Conclusion

Traditional linear programming approaches cannot explain farmers' decisions in regard to the diversification of crops as protection against risk and uncertainty, the augmentation of cereal production with livestock activities to maintain a given degree of self-sufficiency, and the introduction of double cropping activities to solve cash constraint problems.

The informal organization of cooperation between farmers in reducing unused capacity in capital is not conducive to eliminating excess machinery in the village, which results in substantial unused capacity even during peak periods. A formal cooperative to organize the use of machinery, such as tractors and equipment, could reduce costs significantly.

Labor is not a major constraint on farming, if one discounts the population control mechanism in the village that basically results from the concerns of farmers about land fragmentation and leads to a conscious effort to prepare some family members for work outside the village.

In general, farmers appear to be enthusiastic towards new technology, but prefer to observe its results on some other person's farm before trying it themselves. Farmers closely observe experimentation with new techniques or new varieties of seed, fertilizers, and pesticides conducted by 'pioneer' farmers on their own land.

The adoption of new technology is also very much affected by the cash constraints of the farmers. New techniques stand a better chance of being adopted if credit is available to support farmers in their purchase of technology.

The agricultural extension services do not significantly influence farmers in their choice of technology when compared to the impact of fellow farmers. It therefore appears that the successful adoption of new technology critically depends on the availability of credit and demonstration effects observed in the village.

Acknowledgement

We would like to express our gratitude to Mr. Adnan Akcay, Department of Sociology, Middle East Technical University, for his assistance in designing the questionnaire and implementing the village survey.

Implications of Technological Change for Labor and Farming in the Karia Ba Mohamed District, Morocco

DRISS KHROUZ and MOHA MARGHI

Introduction

We have selected the Karia Ba Mohamed (KBM) district for the study of change in farming in relation to labor and technology. The district is located 65 kilometers north of Fez. It comprises 128,000 hectares, of which 92,000 are farmland and 36,000 are pasture, forest, and uncultivated scattered land. Administratively the district is part of the province of Taounate (located in the foothills of the Rif Mountains) and comprises six rural communes.

This district is considered one of the best agricultural zones in the province. The terrain is generally broken, a succession of rounded hills and plateaus intersected with steep-sloped valleys. The subsoil is marly and soils are mostly clay, with a proportion of silt (alluvium) in the valley bottoms. These soils possess a strong capacity for retaining moisture and offer good potential for farming. However, they are subject to erosion on most cultivated sloping land.

The climate is Mediterranean, with wet, cold winters alternating with hot, dry summers. The annual rainfall averages 500 millimeters, 80% of which occurs between November and April. Monthly average temperatures vary between 5° C and 20° C minimum, and 17° C and 40° C maximum.

Private property is the most prevalent (90%), with the remainder of the land owned collectively (6%) or by the state (4%). Most farms in the district are small, as 67% of the farmers own less than five hectares.

The 1982 census reports a population of 130,000 people in the district under study, residing in over 21,000 homes in 447 villages and *douars* (the smallest administrative unit at the level of the rural district). This almost totally rural population depends on agriculture. In the district center, where 3% of the population is concentrated, most branches of public administration are represented by their respective local offices.

The district is an area of rainfed agriculture, with very little irrigation in spite of two major rivers in nearby zones. There are about 1,580 hectares of irrigated land, representing 1.5% of the cultivated area. The principal crops in the district are cereals and legumes, which occupy 85% of the cultivated area. Food crops and livestock predominate in existing systems of production in the district. Technical standards, which are undergoing an important process of change, are still low.

Dennis Tully (ed.), Labor, Employment and Agricultural Development in West Asia and North Africa, 79–102.
© 1990 ICARDA.

The prevailing economy is marked by a predominance of farm family labor, but is also integrated with the market economy both in terms of the purchase of farming inputs and consumer goods, and the sale of produce in rural markets and to middlemen. This integration has been accompanied by other factors, such as (1) the combined effects of value transfer (from rural to urban areas) through price differentials, (2) a population exodus from rural areas to urban centers, and (3) existing subordinate types of work that allow farmers to minimize the effects of fluctuating incomes. Thus, we may say that the district under study is undergoing vast changes as a result of the introduction of new technology as well as other trends that may hamper this process.

We also selected the KBM district for the present study because it has since 1980 been affected by the implementation of a wide-ranging integrated project of agricultural development, partly funded by the World Bank. The Fez-Karia-Tissa Integrated Project is a pioneering experiment in Morocco aimed at the development of areas of rainfed agriculture. These areas, unlike predominantly irrigated zones, had not before 1980 been granted any priority in terms of agricultural investment.

The district has greatly benefited from this project. Its goal is to help increase farmers' incomes, as well as improve their living conditions and general welfare. Project activities are intended to affect not only farming production techniques, but also elements of the educational, medical, and economic environments of the farm population. These activities are expected to narrow the gap between production potentials and the social needs of farmers, which are more and more stimulated by the mass media and by urban consumer models.

The components of the project are the first phase for actions to be undertaken by the government and the private sector in the region. These components concern extension, research, and agricultural training geared toward (1) improving production techniques, (2) making agricultural credit more readily available to small farmers, (3) developing facilities for the supply of agricultural inputs, (4) attempting to reduce soil erosion, (5) establishing a more efficient road network, and (6) setting up schools, medical facilities, and a supply of drinking water in rural areas. A large number of the actions undertaken as part of the project between 1980 and 1987 have occurred in the KBM district.

Accordingly, a pattern of technological change has taken place in the district, demonstrated in the adoption of both farm machinery and inputs such as fertilizer and herbicides. This change has occurred in conjunction with changes in agricultural labor and family needs.

Purpose and Methodology

In this case study, we intend to illuminate the socioeconomic dynamics of rural change through an analysis of the relationship between agricultural labor and technological change.

In undertaking the study, the research team first collected background information on the regional context. We initiated this first phase of the study in three different districts near the city of Fez before deciding to concentrate on the KBM district for the reasons listed above.

Second, we reviewed the literature related to the subject of labor in rural areas (various items of documentation, research project reports, and academic research publications). During this stage, we decided upon the methodological approach to be adopted and the relevant working hypotheses to be formulated.

For the collection of data in the field, we used written survey questionnaires and open-ended interviews with members of the target population. Training was also an essential element of this phase. Two pairs of postgraduate students participated in the survey by recording subjects' answers and taking notes.

The survey sample consisted of 230 farms, selected following a consultation of records at the provincial agricultural statistics office. The purpose was to establish, as adequately as possible, a representative sample. The main criteria for sample selection were size and geographical distribution of farms in the district.

During the last phase of the study, we analyzed the data in 210 of the 215 questionnaires returned (15 farmers declined to participate in the survey) and the notes taken by members of the survey team. We cross-checked the results in order to verify our hypotheses, and compiled this report, taking into consideration the main issues selected for investigation.

General Characteristics of Sample Farms

The general characteristics of farms in the KBM district have determined to a large extent the nature and type of farming practiced there. The most important of these characteristics are related to climate, patterns of land tenure, and systems of production.

One of the most essential features of rainfed farming is that it imposes on the farmer a life characterized by uncertainty, due to the irregular pattern of rainfall, its concentration in one specific period of the year, and sometimes its insufficiency. However, the risks entailed by the weather are attenuated by the favorable soil composition (clay and silt) and other factors that allow farming based on food crops and livestock.

The most important crops grown are cereals and legumes, which occupy 44,200 hectares and 26,000 hectares respectively, amounting to 55% and 32% of the cultivated area. Among cereal plants, soft wheat predominates with 25,000 hectares (56%) while durum wheat covers 39% and barley 4% of the area of cereals. Faba beans are the most important legume crop, followed by chickpeas. Forage cultivation occurs over an area of 4,035 hectares, amounting to 5% of the cultivated area. Commercial crops take up 1,320 hectares with 1,250 for sunflower alone.

About 1,500 hectares are reserved each year for seasonal market crops, particularly cassaba melons and watermelons. Tree crops occupy 11,300 hectares, with olive trees covering approximately 9,700 hectares. In addition 3,400 hectares are left fallow, with the aim of providing grazing, in mos cases after planting has occurred. Starting in December, the unplowed lanc and a few grazing plots are the only feeding grounds for sheep and cattle.

The important stock population comprises 38,200 cattle, 113,000 sheep 6,700 goats, and about 12,500 horses. Farmers practice an intensive type o herding. Animals feed on pasture, grass from unplowed plots, by-product: of crops, stubble, and straw. In quantitative terms, standards of productivit) are still relatively low.

The prevalence of small, excessively fragmented holdings constitutes on(of the most fundamental characteristics of farming in the district. Indeed the total utilized agricultural area for 68% of the farms (owned by 80% o: the farmers in the study) does not exceed 10 hectares. Farms of 10 to 5(hectares represent 28% of the sample (owned by 15% of the farmers), wherea: those with a total utilized agricultural area over 50 hectares make up th(remaining 4% (owned by 2% of farmers). This kind of imbalance also obtain: at the level of land ownership, as only 32% of the farmers own almost 80% of the total utilized agricultural area and 5% of the farmers own 38%. Th(average area per farm (8 hectares) is highly fragmented, with an average o: 12 tiny plots for each farm, each about 0.7 hectare. The situation is ofter aggravated by the long distance between plots, and between holdings anc farmers' dwellings.

Given the broken topography of the area and the inequalities in land tenure any attempt at reorganizing the distribution of holdings is likely to provt a difficult if not impossible task. A consolidation of small holdings, whicł could be an alternative solution and would make the land more accessibl(for the use of machinery, has proved to be unsuccessful. This explains th(slowness of the mechanization process and the reluctance of farmers to accep the agricultural innovations introduced on the larger farms. It also raises th(issue of inequality in the distribution of income and means of production.

Types of Farms

We found three main types of farms in the district, each characterized b) different technological standards and available means of production.

The traditional type includes all of the small farms. Farmers grow mainl) durum wheat, barley, and faba beans. They also raise livestock, but provid(no special care to animals other than meeting their basic needs. The technique: of cultivation are basic; only draft animals are used for the various farminĵ operations. Farmers rely mostly on family labor and make limited use o agricultural inputs. They seldom have recourse to the government-establishec credit facility (*Credit Agricole*) for financing their operations, and most o their production is for subsistence purposes. As incomes are very low anc often not sufficient to cover the families' needs, family members must seel means of complementing them.

The modern type of farm has an area under cultivation that is much larger than the average for the other types of farms. Production of crops is oriented toward the market and is highly diversified, with a predominance of soft wheat, chickpeas, sunflower, and forage crops. Livestock raising is better organized and maintained, with superior cattle breeds used for milk production. Farmers have access to agricultural machinery for various farming operations and regularly use inputs (chemical fertilizers, improved seeds, etc.). They also obtain regular and diversified financing from the government credit institution. Crop yields are generally high (for example, four tons per hectare for cereal crops). Farmers rely on hired labor and farm incomes are high, particularly as some farmers receive extra income from off-farm activities. Thus, this type of farm is able to survive in the face of the risks involved in rainfed agriculture.

The intermediate type of farm shares characteristics with the other two types, but may be distinguished by its heterogeneous nature. There is some diversification of crops, but areas devoted to market crops, particularly soft wheat, chickpeas, and sunflower, are relatively small, compared to those for traditional crops. The mechanization of farming operations is confined to plowing and harvesting, whereas weeding and other maintenance tasks are performed manually. As farmers do not own their equipment, they rent machinery for farming operations. They use some fertilizer and other inputs, but the quantities remain below the amounts required for maximum yields. Expenses incurred during the farming season or for raising livestock with hired labor are commonly financed by the government credit institution. Links with the market are close, but incomes and yields are lower than those for modern farms. The intermediate type of farm is currently undergoing a process of change which reflects the contradictions arising from the use of technology that is not always appropriate for the conditions that may exist at any given time.

Ten percent of the farms in the district are of the modern type, 50% are intermediate, and 40% are traditional. These three types are connected by a series of exchanges, the most important of which are the renting of land and equipment, and the hiring of labor.

Farming Incentive and Support Measures

From the early 1980s, these farms have been affected by a number of interventions on the part of official institutions of agricultural development. For example, the Fez-Karia-Tissa Integrated Project has set up a system of farming extension services in the district, with the purpose of exposing farmers to more effective cultivation techniques, introducing more productive crops, and providing more effective consultation to farmers. Whereas in 1980 there were about 1,800 farmers for each extension agent, the ratio is now 360 to 1. The dissemination of information and advising farmers remain the most important goals of the project. Its methodological approach is based o organizing farmers into service cooperatives for the acquisition and commo use of farm equipment as well as the provision of inputs. The project ha set up an agricultural applied research station for the purpose of helpin solve problems met by farmers in the district.

As is the case in other agricultural regions of Morocco, subsidies are grante in the KBM district for the purchase of inputs and equipment. For exampl the purchase of a tractor and attached equipment is subsidized at rates varyin between 10% and 25%, with preferential rates for farming cooperatives. I the same way, improved fodder seed is subsidized at 40% for individual farmer and 60% for cooperatives. In addition, there is a rate of 40% for weed killer A network of sales outlets for inputs has been established at the level c the rural commune administrative centers, which are for farmers the neares seats of government administration in rural areas. There are also repair shop for farm equipment, together with centers for selling necessary spare parts.

As part of this integrated project, various measures have made agricultur credit for farm operations more accessible to farmers. These include th establishment of local credit offices, the implementation of simplified and mor flexible application procedures, and an increase in financial arrangements base on the actual costs of crop cultivation rather than the total fiscal revenue of farmers, as is the case outside the district.

Finally, we must mention the road network to facilitate the movement o goods and services, and the provision of essential welfare services (schooling health care, drinking water, electrification, etc.) to improve living condition in rural areas.

Changes in Farming

These interventions have had an impact on farming in two important areas namely, technology and the systems of cultivation.

The improvement in technological standards is demonstrated by the observe increase in the number of units of farm equipment and the growing and mor efficient use of agricultural inputs. Indeed, the number of tractors has jumpe from 260 in 1980 to 415 in 1987, an increase of 59%. In 1987, there wa one tractor for every 220 hectares of cultivated land, which amounts to on tractor per 130 hectares if we consider only land accessible to machinery This is near to the ratio reported for large irrigated areas, where mechanizatio is much more important.

The number of threshing machines has been multiplied by six (26 in 198 versus 154 in 1987). The same increase has occurred in the case of equipmen used for plowing, sowing, and mowing. In addition, the use of farming input has grown about 45% over the same period, and farmers apply these product more and more efficiently. We must point out that these achievements ar relevant only in the case of modern farms and a significant proportion o intermediate farms.

There have been similar effects on the systems of cultivation, especially in regard to crop substitution, cultivation techniques, and crop yields.

In the case of cereals cultivation, there has been an increase in the areas devoted to soft wheat at the expense of more traditional cereal crops (barley and durum wheat). Overall, cereals still remain important, and the area occupied by soft wheat (11,500 hectares in 1980) has increased to 22,300 hectares in 1987. We may explain this increase by the market value of soft wheat, which is subsidized at the price of 2,000 dirhams (MAD) per ton (1 USD = 8 MAD), and by its greater productivity, which depends on an effective use of technology.

Turning to legumes, we note a slight reduction in the areas for faba bean and lentil crops, whereas holdings planted in chickpeas have significantly increased (3,700 hectares in 1987, compared to 1,200 in 1980).

The cultivation of fodder crops has increased in a similar fashion (4,000 hectares in 1987, compared to only 1,100 in 1980). We should also mention that medic, in rotation with cereal crops, has been introduced in the district and is expanding into other regions of the country.

The cultivation of sunflower is fully established. In 1980, it occupied only 230 hectares, compared to about 1,800 hectares in 1987. In spite of crop losses caused by sparrow invasions, farmers have benefited from the cultivation of sunflower because of the current subsidized price (4,150 MAD per ton in 1987, as opposed to only 1,800 in 1984), mechanization, guaranteed marketing, and its place as a substitution crop at the end of the cycle.

Fallow fields are fewer in number, particularly because of the introduction of machinery that allows the farmer, formerly constrained by the farming calendar, to sow his land more quickly. However, a small proportion (5%) of the fallow land is still maintained for grazing.

Generally speaking, soil preparation, which is an essential stage in the planting of crops, has been the area most affected by the use of technology. This constitutes an important factor in the determination of crop yields in rainfed farming. Indeed, soil preparation now occurs during the dry conditions of summer, an achievement that would be impossible for clay and silt soils without the introduction of powerful machinery.

The improvement of cultivation techniques varies according to farming operation and type of crop. As far as cereal crops are concerned, we have already mentioned significant improvement in the most important operations that are mechanized, namely, plowing, planting, and harvesting. On the other hand, mechanization has hardly affected maintenance operations (weeding and fertilization). Farmers are reluctant to use machinery for these operations because the equipment is likely to damage the crop. In the case of legumes and sunflower, farmers use drills for planting, leaving spaces wide enough for weeding and other maintenance tasks.

We may also consider the practice of crop diversification and the techniques associated with it as an important technological development that contributes towards the improvement of farmers' incomes.

In addition, crop yields are a valid criterion for evaluating the results of attempts to introduce farming technology in the district. An analysis of the survey data reveals that the average yield for the main cereal crops has increased from 150 kilograms per hectare in 1980 to 210 in 1986 (two years with comparable weather conditions). The average faba bean yields for the same years are 70 and 100 kilograms per hectare. These examples show a general tendency toward an increase in the productivity of major crops in the district.

Our analysis has so far revealed the heterogeneity that exists in the KBM district, in regard to land distribution, types of farm, and production systems. This explains the fact that changes introduced through the adoption of technology in the district have had two contradictory effects: on the one hand, the modernization of farming and higher incomes on modern farms and, on the other, an almost complete stagnation of production on traditional farms, with an exclusive reliance on manual labor by family members who seek extra income by working for wages on the mechanized farms.

This analysis raises the fundamental issue of the kind of interaction that obtains between mechanization and labor patterns in agriculture.

General Characteristics of Agricultural Labor

Agricultural labor in the district, where rainfed farming and small holdings prevail, is characterized by complexity. In this respect, labor is an important element in the process of change created by the introduction of technology.

This complexity is first demonstrated in the diversity both in the supply of labor and the kinds of people involved in the search for wage-earning work. In most cases, all family members are available for work, and a division of labor exists within the family according to age, gender, and state of health, where each member is assigned a task or activity.

The second feature of agricultural labor is the irregularity of demand, in spite of the variety of jobs offered and tasks to be performed. This demand is closely related to the timetable of farming operations, which is itself determined by the prevailing types of crop production and animal husbandry, and the influence of the climate. Therefore, an employer's labor requirements will frequently depend upon the abundance or lack of rain.

A third equally important feature of agricultural labor in the district is the distribution of regular versus occasional labor, and family versus wage labor, which gives rise to a situation of diverse relationships conditioned by the type of farming. Part-time labor is more common in this context; permanent labor is found on large farms that employ individuals with specific quali-fications.

In the KBM district, the labor market includes four fairly distinct categories of people available to work for wages. The first category includes landless laborers who live in the area and who are attached to certain farms near the villages where they live. Wage labor is the only way for members of this category to earn a living. The second category consists of small farmers who work relatively small plots of land in cooperation with members of their families. Once they have completed their own farming, they work for wages on large and medium-sized farms to supplement the income they derive from their own land.

The third category includes workers from neighboring or far-away districts who come to the KBM area for the purpose of finding work, not necessarily agricultural. The last category is made up of individuals (from any of the first three categories) who demonstrate more personal initiative by renting or sharing plots of land for cultivation.

In farm families all persons old enough to work are available for employment either on the farm itself or outside. In the sample surveyed for the case study (where the mean size of farms was eight hectares), we found an average of nine persons per farm. Women, whose labor in the past was restricted to housework and work on the farm, now work for wages. Similarly, children leave school to help herd stock and do other odd jobs for pay. Family labor is most important on small and medium-sized farms, with an average of four persons (approximately three manpower units) involved in agricultural work.

An average farm with eight hectares and four large animal equivalents requires about eighty paid working days, in addition to family labor. Indeed, if we take into consideration the number of available manpower units (three per farm) and the number of days available for farming operations (around 180 in the KBM district), we notice that hired labor exceeds 25% of the total. However, the importance of hired labor varies according to the type of farm.

On the modern, mechanized farms, the total utilized agricultural area is of considerable size, but the amount of hired labor is not correspondingly large. Family labor is devoted to other tasks and farm machinery enables work of a good standard to be completed on schedule.

In the case of traditional farms with smaller cultivated areas, family labor is more important. The relationship between hired labor and total utilized agricultural area is relatively significant, mainly due to the cultivation techniques used on this type of farm, from planting through harvesting. Manual labor is predominant. Farmers frequently resort to hired labor during peak periods or when cultivated areas are large enough to require extra workers.

Agricultural workers in the district are either full-time ('permanent') or part-time ('occasional') workers. The former are workers whose regular presence on the farm is necessary because of the tasks assigned to them. They herd livestock, watch over crops in the fields, drive tractors, and supervise other workers. Generally over thirty-five years of age, they receive more or less stable wages (20 MAD per day) throughout the year. Sometimes they receive payment in kind (for example, a sheep for a Muslim holiday, wheat, butter, milk, olive oil). These workers, exclusively male, are found on the large farms and represent less than 3% of wage labor in the district. Hiring is not easy and is generally based on the employer's trust. Employers frequently ask such workers to perform duties different from the ones for which they were hired. Thus, the tractor driver becomes a mechanic or a handyman inside the farm house, and may be required to help his employer with the sale of farm produce at the market. On farms where most workers are permanent, there is one worker per thirty-eight hectares for crop cultivation and one per twenty-seven hectares for crop cultivation and livestock raising.

Farmers hire part-time ('occasional') workers only during certain periods when their needs for labor are high. The hiring varies according to the time of the year and the type of farm. This category comprises workers of all ages, including women and young girls. Individuals in this category receive the lowest wages, normally 10 to 15 MAD per day. However, these wages may go up to 40 to 50 MAD per day for adult males during busy periods such as harvest time.

Farm employers recruit part-time agricultural workers, who make up 9% of the agricultural labor force in the district, from the category of landless laborers, according to the immediate needs of the farm and for very short periods, sometimes just for half a day or even a single produce-loading operation. The hiring process, which is often the responsibility of full-time workers rather than the prospective employer himself, favors youths and women for certain operations, with the purpose of keeping down costs.

There is an abundance of part-time labor, as the district receives workers from neighboring regions, particularly for picking olives, hoeing legumes, and harvesting cereal crops. On the other hand, during slack periods in the farming calendar workers from the KBM district move to towns in search of non-agricultural employment.

The Agricultural Calendar and Demand for Labor

When we look at the agricultural calendar, we find that the demand for labor is concentrated during certain peak periods (Figure 1). Thus, the demand for part-time workers is at its highest during the cereal crop harvest in June. Once wheat has matured around the end of May, there is a risk of thunderstorms that may damage the crop if it is not harvested quickly enough. Moreover, relatively few threshing machines are available at this time. Finally, farmers' credit payments are due at the beginning of July.

All of these factors make the demand for part-time harvest workers particularly acute. During other periods of the year, the need is less urgent. Plowing, which takes place in summer on the mechanized farms, begins only after the first autumn rains (the end of October) in the unmechanized sector. On these farms, the operations of plowing, fertilization, and spreading seed occur during the period between October and December. However, due to the nature of the soil (clay and silt) and the concentration of rainfall, part-time labor is often hired in order to complete planting on time. This is also the period when legumes and forage crops have to be planted. November is a peak period for the olive harvest and sale, which must take place

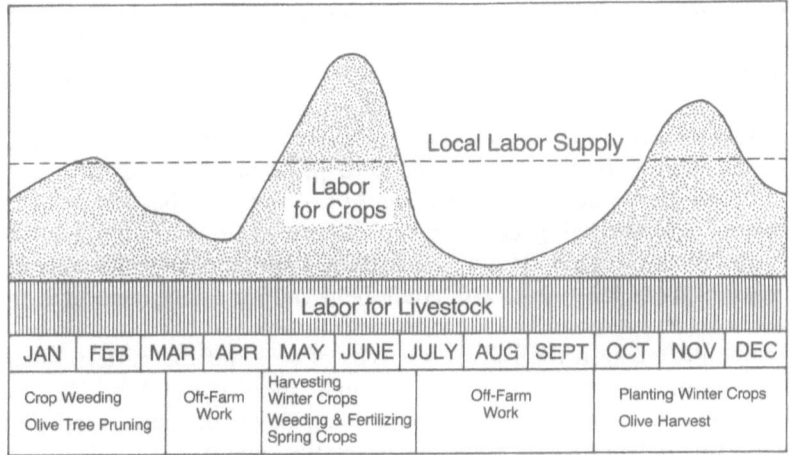

Figure 1. Labor Supply and Demand in Karia Ba Mohamed District.

simultaneously to avoid losses in oil.

January and February are the months for weeding legumes and cereal crops, applying nitrate fertilizers, plowing to prepare for spring crops, and pruning olive trees. The demand for hired labor tends to fluctuate according to the amount of rain in winter that causes the growth of weeds. The degree of soil moisture and the growth rate of the crops also affect the choice between farm machinery and hired labor by owners of large farms.

The period between March and April is less busy, as the operations affect only certain farms in the modern sector. These operations include olive tree care, planting sunflowers and chickpeas, and the cultivation of seasonal market crops. May and June, on the other hand, constitute a peak period when demand for wage labor is high. Hoeing and thinning spring crops, and harvesting legumes, fodder, and (especially) cereal crops, are the most important operations that cause a marked increase in the demand for labor. This high demand is met by the arrival of many part-time workers from neighboring districts. Daily wages are relatively high, especially in the case of cereal crop harvesting performed under quite difficult working conditions.

Activities for which there is less need for hired labor occur throughout the year. They include herding animals and cultivation of forage crops, which do not require any particular attention between sowing and harvesting.

The Impact of Mechanization on Agricultural Labor

As we have seen, agricultural labor is seasonal and demand is greatest during the short periods for plowing, planting, weeding, and harvesting. Moreover, the development of mechanization in the KBM area has greatly affected not only systems of cultivation but also, and as a consequence, the size and organization of agricultural labor.

As far as large and medium-sized farms are concerned, mechanization allows plowing and planting large areas to take place over fairly short periods of time. Whereas with animal traction, currently used on small farms, a farmer needs an average of six to seven days to plow one hectare, with a tractor he can plow the same area in only two hours and do a better job. Similarly, harvesting and threshing cereal crops require eight workdays on farms of the traditional type, on medium-sized farms situated in rough terrain, and on small plots of land that farm equipment cannot easily reach. On modern farms, however, combines can harvest and thresh eight hectares in one day.

We should mention that, given the type of soil and the pattern of rainfall, the use of mechanized equipment on small plots of land is risky because the movements of heavy machinery pressing down the soil, resulting in the deterioration of soil structure, may have negative repercussions on crop growth. On these small farm holdings, farmers resort to manual labor.

The most important effects of mechanization on agricultural labor are related to (1) the qualifications of workers, (2) the organization of labor, and (3) the prevailing wage levels.

In regard to the qualifications of workers, mechanization in the district has resulted in the creation of jobs for workers with specific skills, particularly truck drivers and tractor operators. These jobs are generally held by permanent workers. Similarly, there is a demand for mechanics and welders, jobs that require new types of qualifications that extend beyond the mere relationship between man and machine and have important implications for the quality of mechanized operations. Workers must know how to contribute their skills toward the improvement of cultivation techniques. Labor recruitment thus implies a strict selection of qualified individuals with the ability to demonstrate knowledge of the machine, the tool, and the cultivation operation. Operators of modern farms will hire individuals with a wide range of qualifications, obviously for purposes of reducing the number of workers and thus cutting down costs. For example, a hired tractor driver must necessarily be able to adjust the plow for the most effective soil preparation, set the seed drill for a given quantity of seed, and regulate the sprayer for a specific dose of fertilizer. In addition, he will be called upon to carry out maintenance and repair work on various items of equipment.

Mechanization has also affected the organization of agricultural labor in the district. The data from our sample show that, in cases where mechanization standards are high, particularly on large farms, there is first an allocation of new tasks and assignments to family labor available on the farm. Family members are exempt from cultivation, and assume responsibility for the purchase of farm supplies and the sale of produce. Second, there has been the creation of new skilled occupations such as tractor drivers and mechanics. Farmers who rely on farm equipment tend not to hire part-time workers except when absolutely needed. Still, neither the introduction of farming technology nor the mechanization of cultivation techniques has managed to totally replace part-time wage labor on large and medium-sized farms. Instead, there has been a change in its structural pattern, a transfer of labor from one operation to another, as a consequence of the crop diversification resulting from mechanization.

Turning to wage levels, we must first point out that existing labor legislation has established a guaranteed minimum wage in agriculture. Theoretically, one of the main reasons for this action was to determine a minimum income per working day (eight to nine hours according to the season) for the least qualified agricultural workers, who would be entitled to extra pay for any additional labor. In 1973, this guaranteed minimum wage was 6 MAD per day; it increased to 10 in 1980 and at present is approximately 20 MAD per day.

The wages offered to agricultural workers vary considerably, even within the same area, according to the time of year, the tasks involved, and the age and gender of the worker. These wages are not in any way indexed to the cost of living. For example, between 1980 and 1987, the guaranteed minimum agricultural wage has hardly doubled, whereas the cost-of-living index has gone up two-and-a-half times.

Workers belonging to the full-time ('permanent') category seem to be at an advantage regarding wages, although the wage may vary according to the job. Thus, in the KBM district, the average daily wage of a shepherd is 12 MAD, a cowherd receives 15, a field guard 17, and a tractor driver 23.

In spite of additional advantages in kind received by some workers, wages remain relatively low, particularly for those working in the livestock sector. As herders need no specific qualifications, these workers can be easily replaced, which makes their situation most unstable. During peak periods, full-time workers sometimes receive one or two dirhams extra, depending on their employer's disposition. They also avail themselves of the opportunity to recruit members of their families as part-time workers, a practice that to some extent helps increase their incomes. The most important advantage members of this category enjoy, compared to part-time workers, is the guarantee of almost certain wages (however low they may be) throughout the year.

Part-time workers are mainly youths and women. The wages they receive are generally below the guaranteed minimum wage and tend to vary according to age, gender, and the kind of farming operation involved. The daily wages vary between 10 and 20 MAD during normal periods, and may reach 30 to 50 MAD for male adults during peak periods (olive picking, cereals harvesting, etc.). Harvesting legumes, manual weeding, and hoeing sunflower and chickpea crops are the operations for which the lowest wages are paid.

The mechanization currently taking place in the KBM district has resulted in a reduction in the time needed for farming operations on modern farms. Such mechanization has brought about a need for qualified laborers whose ability to use tools and machinery is likely to contribute to the improvement of cultivation techniques and production. Because they have the right qualifications, and skilled workers are not readily available on the labor market, tractor drivers, operators of other farm machines, mechanics, and welders often demand wages higher than the guaranteed minimum agricultural wage. This means that employers offer wages that have been more or less negotiated in order to complete their farming operations within the required time limits. The owners of mechanized farms generally try to keep in close contact with as small a contingent of part-time workers as possible, in addition to full-time ones.

Full-time workers with specific qualifications related to farm machinery receive the highest agricultural wages in the district. In addition to wages averaging 23 MAD per day, they receive bonuses during peak periods and other advantages in kind. For part-time workers, daily wages are lower than the guaranteed minimum agricultural wage for at least six months of the year.

The relative increase in wages justified by workers' qualifications shows the direct influence of mechanization on incomes. Moreover, a number of support activities and services have appeared to address the new needs created by mechanization. Thus, sales shops for farm machinery have opened and a network of spare parts trading is in operation. Similarly, there are workshops for maintaining and repairing farm machinery. These activities create new jobs for local labor in establishments that provide services at fairly competitive costs. Most repairs take place on the spot, whereas only a few years ago the mere cost, in both time and money, to transport a damaged piece of machinery into town for repairs amounted to many times the present cost of a spare part purchased locally. However, these activities are restricted by the limited availability of financing and the very close links between farmers of the mechanized sector and neighboring urban centers.

Crop Cultivation Costs and Agricultural Labor

In the KBM district, there has been a substantial increase in crop cultivation costs, resulting from the mechanization of cultivation techniques and the use of inputs, which represent essential factors in achieving high crop yields. The share of costs devoted to labor varies according to the crop and the type of farm. Generally speaking, the overall costs per hectare under cultivation are higher for mechanized farms, and the share taken up by labor fairly small.

We analyzed the costs and labor expenses on the three types of farms (traditional, intermediate, modern) for the main crops in the district, namely soft wheat, faba beans, sunflower, forage crops, and olive trees.

Expenses for one cultivated hectare of soft wheat are three times higher in the case of modern mechanized farms than they are for farms in the traditional sector (Table 1). On traditional farms, labor expenses amount to approximately 40% of the costs for one cultivated hectare because farmers resort to hired labor in addition to family labor for planting, weeding, fertilization, and

Table 1. Inputs of Soft Wheat Production.

Farm Operations	Number of Work Days	Machinery Cost (MAD)	Amount of Inputs	Cost of Inputs (MAD)
Traditional Farms				
Prepare seed bed (animal traction)	4.0			
Fertilization: deep	.5		50 kg	80
surface	–		25 kg	35
Planting (local seed)	1.0		150 kg	300
Weed control: Antimono	3.0		1 l	30
Harvesting (manual)	6.0			
Threshing	3.0			
Transport	2.0			
Total	19.5			445
	(4 with animals)			
Semimechanized Farms				
Prepare seed bed: medium tillage		200		
cover crop		90		
Fertilization: deep	.5		150 kg	240
surface	.5		50 kg	75
Planting (improved seed)	.5	90	170 kg	425
Weed control: Antimono	1.0		2 l	180
Antidico	1.0		1 l	30
Crop protection	15.0			
Harvesting (mechanical)		250		
Trussing		200		
Transport				
Miscellaneous		100		
Total	18.5	930		950
Mechanized Farms				
Prepare seed bed: medium tillage		200		
cover crop		270		
Fertilization: deep	.5	60	200 kg	320
surface	.5		100 kg	150
Planting (improved seed)	.5	90	170 kg	425
Weed control: Antimono	1.0		3 l	270
Antidico	1.0		1 l	30
Antiparasite	1.0		.5 l	260
Crop protection	10.0			
Harvesting (mechanical)		250		
Trussing		220		
Transport		50		
Miscellaneous		100		
Total	14.5	1240		1455

SOFT WHEAT PRODUCTION

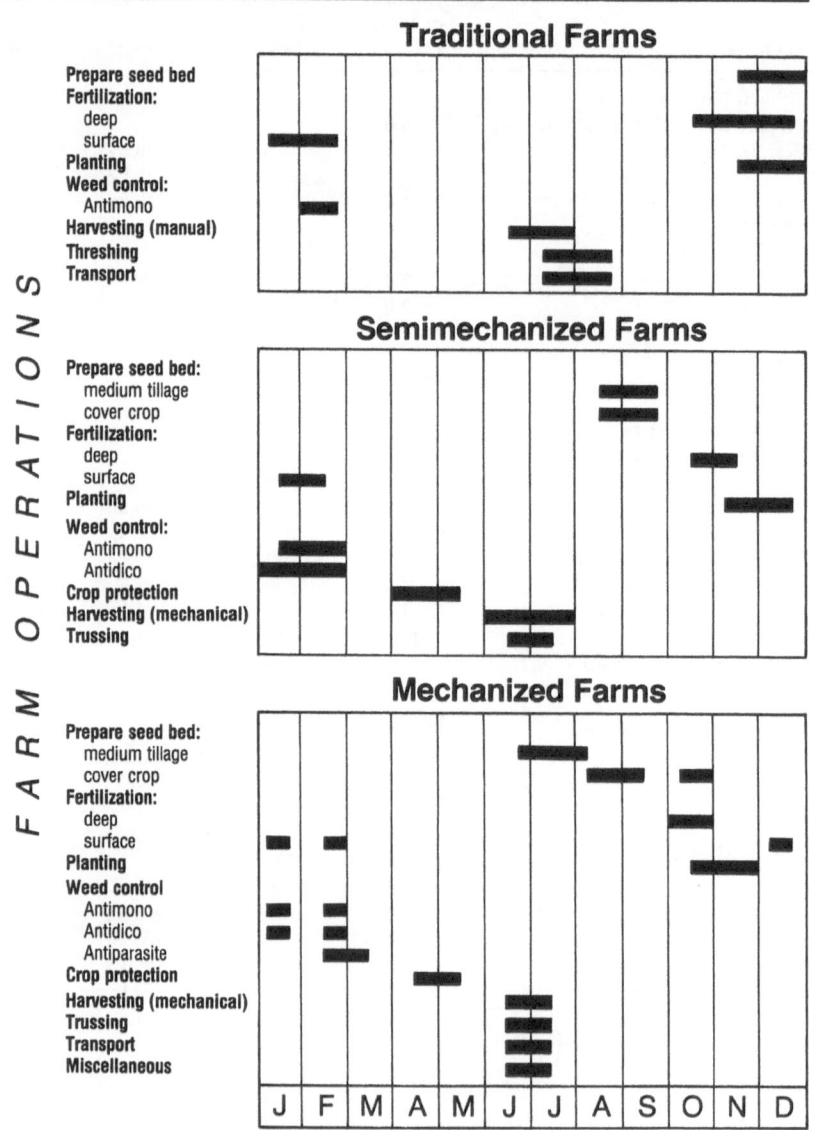

Figure 2. Soft Wheat Production on Traditional, Intermediate (Semimechanized), and Modern (Mechanized) Farms: Schedule of Operations.

harvesting. Harvesting is the most costly operation in this respect. Raising funds for harvesting can be problematic if crop yields are low due to bad weather at the end of the agricultural cycle (Figure 2).

For one hectare of soft wheat in intermediate or semimechanized farms, expenditure on labor makes up 15 to 20% of the overall costs per hectare. Farmers rent equipment for plowing, whereas they rely on seasonal manual labor for planting, weeding, and fertilization. They often hire equipment either from contractors in neighboring districts or from farmers in the mechanized sector who have completed their own farming operations. In fact, renting out equipment constitutes a supplementary source of income for farmers who own machinery, and may encourage others to purchase their own.

In modern farms, on the other hand, all cultivation operations are mechanized, and labor expenses are figured separately, together with input supplies. They amount to 10-15% of total costs. On this type of farm, there is almost always a contingent of permanent labor, with important skills and correspondingly high wages.

Table 2. Inputs of Sunflower Production.

Farm Operations	Number of Work Days	Machinery Cost (MAD)	Amount of Inputs	Cost of Inputs (MAD)
Semimechanized Farms				
Prepare seed bed: medium tillage		200		
cover crop		90		
Fertilization: deep		60	150 kg	240
surface	.5		50 kg	75
Planting	1.0		9 kg	90
Ridging	4.0			
Hoeing	10.0			
2nd Hoeing	2.0			
Picking	10.0			
Miscellaneous		250		
Total	27.5	600		405
Mechanized Farms				
Prepare seed bed: medium tillage		200		
cover crop		270		
Fertilization: deep		60	250 kg	400
surface			50 kg	75
Planting		90	9 kg	90
Hoeing (manual)	10.0			
2nd Hoeing (mechanical)	4.0	100		
Crop protection	15.0			
Picking	10.0			
Miscellaneous		250		
Total	39.0	970		565

96

We have already mentioned that sunflower is a fairly recent crop in the district. The pattern of labor presents striking differences with that for other crops, as sunflower cultivation is restricted to mechanized and semimechanized farms (Table 2). Labor expenses are higher than those recorded for the cultivation of cereal crops because of the greater number of operations that require labor. For example, hoeing may have to be repeated at least three times, depending on how heavy the rainfall is in March and April (Figure 3). Battling against sparrow invasions and harvesting the crop are also tasks that require labor for sunflower cultivation. For purposes of comparison, our data show that labor expenses amount to 39% and 42% of cultivation costs in the mechanized and semimechanized sectors respectively. The results also show that for one cultivated hectare of sunflower on farms of the intermediate type, at least twenty-seven days of wage labor are needed. For the mechanized type, this number reaches thirty-nine days of part-time wage labor, excluding permanent labor used on this type of farm.

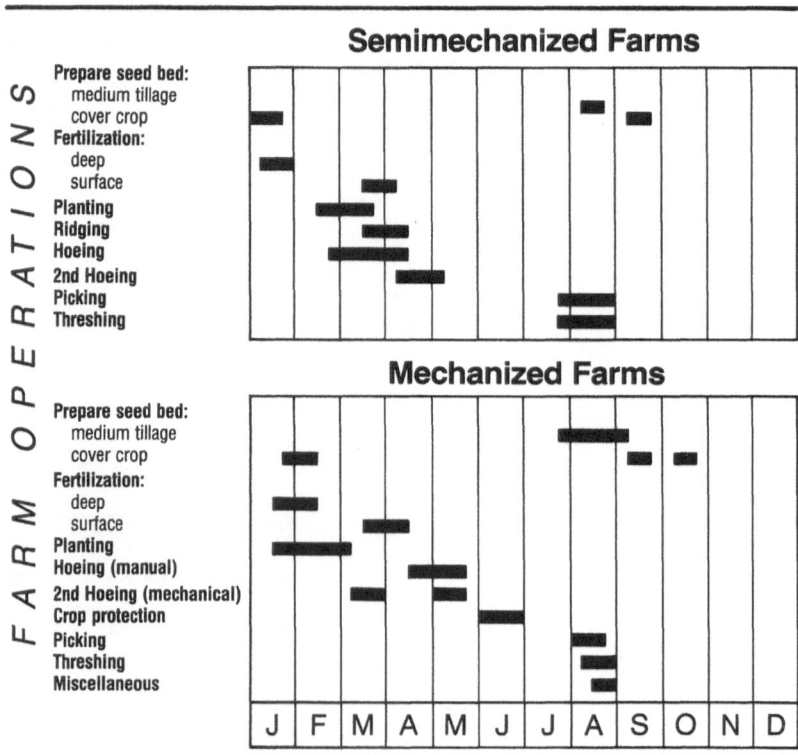

Figure 3. Sunflower Production on Intermediate (Semimechanized) and Modern (Mechanized) Farms: Schedule of Operations.

Table 3. Inputs of Faba Bean Production.

Farm Operations	Number of Work Days	Machinery Cost (MAD)	Amount of Inputs	Cost of Inputs (MAD)
Traditional Farms				
Prepare seed bed (animal traction)	4			
Fertilization	2		20 kg	120
Planting	2		100 kg	250
Hoeing	4			
Manual harvest	8			
Transport (animal)	3			
Threshing	1			
Total	24			370
Semimechanized Farms				
Prepare seedbed: medium tillage		200		
cover crop		90		
Fertilization	2		200 kg	120
Planting	2		100 kg	250
Weed control: Antimono	3		1 l	250
Hoeing	2			
2nd Hoeing	2			
Manual harvest	7			
Transport	2			
Threshing	1	100		
Total	21	390		620
Mechanized Farms				
Prepare seedbed: medium tillage		200		
cover crop		180		
Fertilization	2		400 kg	240
Planting	2		100 kg	250
Weed control: Antimono	3		1 l	250
Hoeing	4			
2nd Hoeing	4			
Manual harvest	7			
Transport	1			
Threshing		200		
Total	23	580		740

Because of the type of operations involved, the cultivation of faba bean crops is similar to that of sunflower, hence their comparability in terms of the use of labor. Therefore, mechanized and intermediate farms exhibit more or less the same pattern concerning the number of wage labor days that average twenty per hectare, representing approximately 25% of expenses. The traditional sector, which resorts to family and occasional labor for all cultivation operations, is characterized by much higher rates, with 38% of the total costs

98

devoted to labor (Table 3, Figure 4).

Forage crops need very little labor, as all operations are mechanized (Figure 5). The number of wage labor days per hectare is the lowest recorded, and labor expenses amount to less than 10% of the total cultivation costs (Table 4). This is due essentially to the nature of this type of crop, which is meant for animal consumption. The mechanized cultivation of these crops requires no herbicide treatment or weeding.

The cultivation of olive trees is the only agricultural endeavor in the district whose operations, performed by ordinary laborers, are not yet mechanized

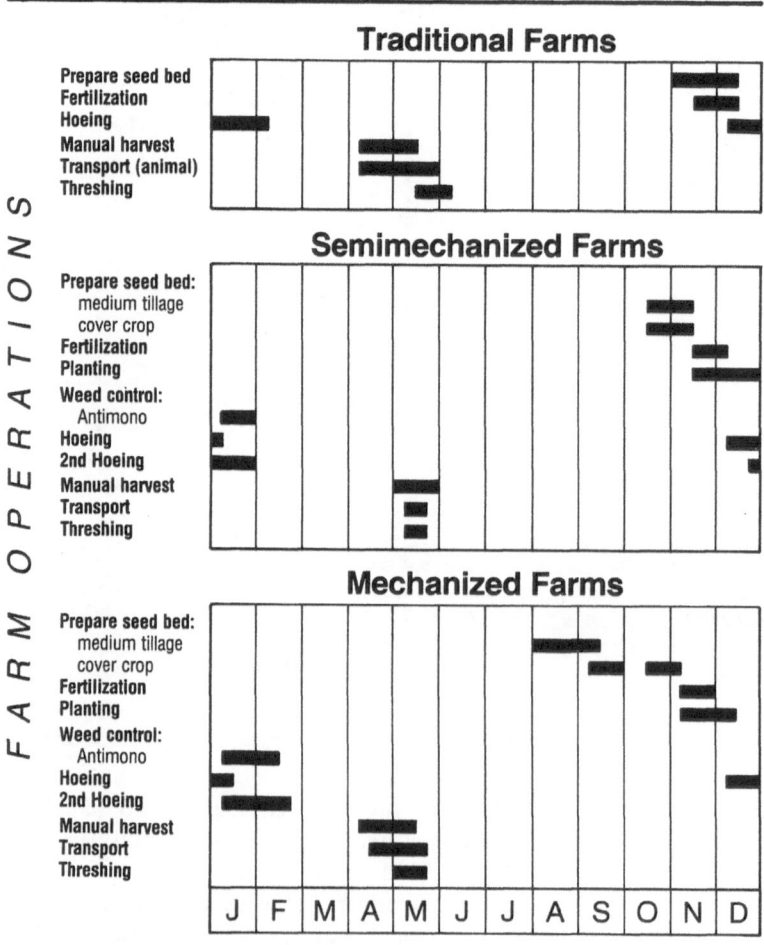

Figure 4. Faba Bean Production on Traditional, Intermediate (Semimechanized), and Modern (Mechanized) Farms: Schedule of Operations.

FORAGE CROP PRODUCTION

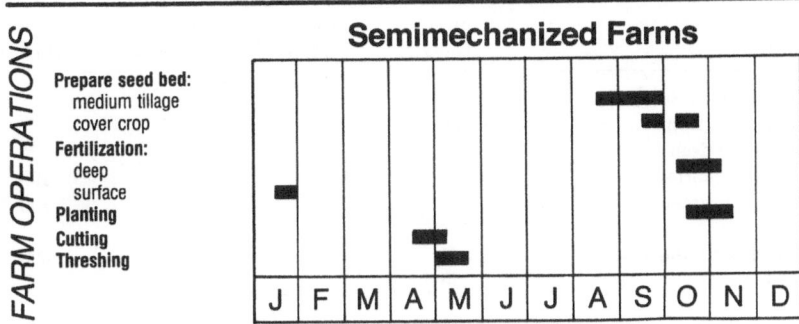

OLIVE PRODUCTION

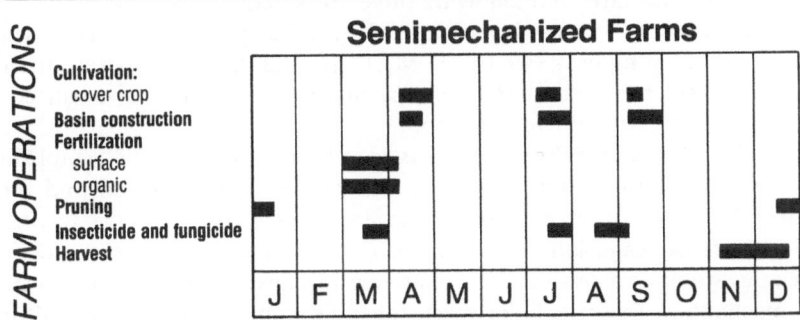

Figure 5. Forage Crop and Olive Production on Intermediate (Semimechanized) Farms: Schedule of Operations.

Table 4. Inputs of Forage Crop Production on Semimechanized Farms.

Farm Operations	Number of Work Days	Machinery Cost (MAD)	Amount of Inputs	Cost of Inputs (MAD)
Prepare seedbed: medium tillage		200		
cover crop		180		
Fertilization: deep	.5		200 kg	300
surface	.5		50 kg	75
Planting	.5	90	120 kg	400
Cutting		60		
Threshing	1.0	350		
Transport	1.0	80		
Total	3.5	960		775

Table 5. Inputs of Olive Production on Semimechanized Farms.

Farm Operations	Number of Work Days	Machinery Cost (MAD)	Amount of Inputs	Cost of Inputs (MAD)
Cultivation by cover crop		270		
Basin construction	9			
Fertilization: deep	2		300 kg	375
surface			100 kg	100
organic	5		1 ton	100
Pruning	8			
Insecticide and fungicide	4			600
Harvest	30			
Total	58	270		1175

(Figure 5). One hectare planted with olive trees requires about sixty wage labor days (Table 5). Harvesting olives, which takes place at a period when wages are at their highest (40 to 50 MAD per day), is an operation that requires a considerable amount of labor. Sixty percent of cultivation costs per planted hectare are allocated to labor.

Our analysis of crop cultivation costs in relation to labor has established the importance of the latter, both in terms of the costs involved and their place in relation to other expenses incurred by farmers. We have recorded that the share of wage labor in cultivation costs decreases with the degree of mechanization, as a result of the relative awareness, on the part of farmers in the mechanized sector, of the work schedule imposed by the agricultural calendar. Furthermore, mechanization is accompanied not only by an increase in crop cultivation costs, but also by substantially higher yields (Table 6). Finally, crop diversification has particularly favored the crops for which the demand in labor is high.

Conclusions

This case study of the KBM district is meant as a contribution to research on agricultural labor and technological change in rainfed farming areas. The district is characterized by positive agricultural potentialities because of its soils and rainfall patterns. Since 1980, it has been the target of a policy of deliberate, supervised agricultural development. This policy has induced substantial changes in agriculture, in spite of strongly constraining agrarian land distribution and ownership patterns. Indeed, mechanization has caused a shift from a system of production based mainly on a few food crops toward an intensification of agriculture and a system of more diversified crop production and livestock raising. This system provides greater security to

Table 6. Economics of Crop Production (MAD per ha).

	Soft Wheat			Sunflower	
	Trad[a]	Semi[b]	Mech[c]	Semi	Mech
Average yield (kg)	1100	2500	3500	1200	1800
Value of yield	2105	4800	6675	4920	7380
Cost of inputs	445	950	1455	405	565
Cost of labor	400	450	475	690	975
Cost of draft animals	140	–	–	–	–
Cost of machinery	–	930	1240	600	970
Net profit	1120	2470	3505	3225	4870

	Faba Bean			Forage Crop	Olives
	Trad	Semi	Mech	Semi	Semi
Average yield (kg)	800	1200	1600	4500	50 kg/tree
Value of yield	1600	2300	3200	4500	10000
Cost of inputs	370	620	740	775	1175
Cost of labor	425	500	575	100	1450
Cost of draft animals	200	–	–	–	0
Cost of machinery	0	390	580	960	270
Net profit	605	790	1305	2665	7105

[a] Traditional farms
[b] Semimechanized farms
[c] Mechanized farms

farmers in facing the uncertainties of rainfed agriculture. The transformation from a peasant economy to a market-oriented one is well under way. However, the preponderance of small farms, with all their constraints, contributes to the phenomenon of resistance to technological progress in rural areas.

On the other hand, agricultural labor is closely linked to systems of production and cannot be excluded from the process of change that is taking place. The results of this study have suggested that some current opinions about labor and mechanization, often derived from preconceived ideas, need to be reconsidered. Thus, mechanization, which has often been considered a cause of unemployment or underemployment, has in this case brought about an intensification of cultivation techniques and a diversification of crops. As a result, there has been neither unemployment nor underemployment, but rather the creation of permanent jobs with higher qualifications and a redistribution of tasks between available family labor and wage labor. Agricultural labor has shifted from mechanized crops to new crops requiring more labor which is generally available and more expensive outside peak periods. Crop diversification has created job opportunities that exist throughout the agricultural year.

Moreover, the technical support of mechanization has indirectly contributed to new rural job opportunities. Mechanization has also brought about an increase in productivity and an improvement in crop yields, as a result of an effective mastery of cultivation techniques and a better use of inputs, contributing toward a relative increase in wages and to the improvement of incomes.

As a follow-up to the present study, further research is needed to address the problems faced by farmers in the traditional sector, particularly the need to set up new forms of organization, and a better use of available labor suited to crops and agricultural schedules.

Specifically, there are several issues that seem to deserve further attention on the part of researchers. First, they should investigate the possibility of introducing small-scale mechanization suited to farms with reduced areas for cultivation. There is also the existing imbalance in the distribution of land that constitutes the most important constraint to the adoption of short or medium-term solutions to the problems of small farms. Research is also needed that would help find techniques suited to the cultivation of different crops in rainfed agricultural areas, and models of farming that could be adopted for this kind of agriculture, especially to minimize the risks due to uncertain rainfall.

Finally, researchers should reflect on ways to strengthen the mechanization process in agriculture for both small and medium-sized farms, and to create local-level facilities for processing agricultural products, which would most certainly help toward an improvement of the prevailing labor situation.

Mechanization and Agricultural Employment in Arid and Semiarid Zones of Morocco: The Case of Upper Chaouia

LARBI ZAGDOUNI and DRISS BENATYA

Issues and Objectives of the Study

During the last fifteen years, the mechanization of agriculture in Morocco has progressed more rapidly than previously, although the amount and types of equipment are still judged to be inadequate. As one of the technical levers for producing a much-desired increase in cereals for consumption, agricultural mechanization benefits from a number of measures implemented by the state, such as exempting certain equipment from import taxes, financing by the government credit institution (*Credit Agricole*), and granting crop subsidies.

From the producers' point of view, mechanization serves a variety of purposes. For example, entrepreneurs perceive the acquisition of farm machinery for contract operations as an opportunity for realizing a greater return on their investment. The motive for adopting mechanization may be the search for lower labor costs or an increase in labor productivity. The use of agricultural machinery can be rationalized as being less expensive than traditional tools and methods, or seen as a way of reorganizing the management of resources for undertaking new activities. In other words, the present spread of agricultural mechanization is part of a highly complex process in which greater productivity of the land is not necessarily the only goal.

The rapid growth of mechanization bears investigating, particularly as it stands in marked contrast to the very meager advances made in productivity. Does this situation indicate a deep disharmony between the choices of the state and the strategies of producers with respect to the goals of agricultural mechanization? This question becomes even more pertinent when one takes into account the changes that are occurring in the realm of agricultural policy. For instance, the arid and semiarid zones have recently received greater attention from policy makers who hope to increase the productivity of cereals and reduce shortfalls. Among the most favored means is small-scale mechanization; 70,000 to 80,000 small tractors will be introduced in the next twenty years. In addition, increasing employment in rural areas and improving the general quality of life there are extremely important in the effort to slow rural flight towards urban centers.

In the particular case of arid and semiarid zones, where the possibility of diversifying and intensifying production is after all quite limited, does this

Dennis Tully (ed.), Labor, Employment and Agricultural Development in West Asia and North Africa, 103–140.
© 1990 ICARDA.

emphasis on mechanization contradict the goal of promoting agricultural employment?

The Study Area

The Upper Chaouia region in the province of Settat is one of the largest grain-producing areas in the country. Family agriculture is still preponderant here. The choice of this area was based particularly on the fact that for a decade it has been the object of numerous research projects. Their results have shown that farm mechanization, which is of recent origin and is progressing

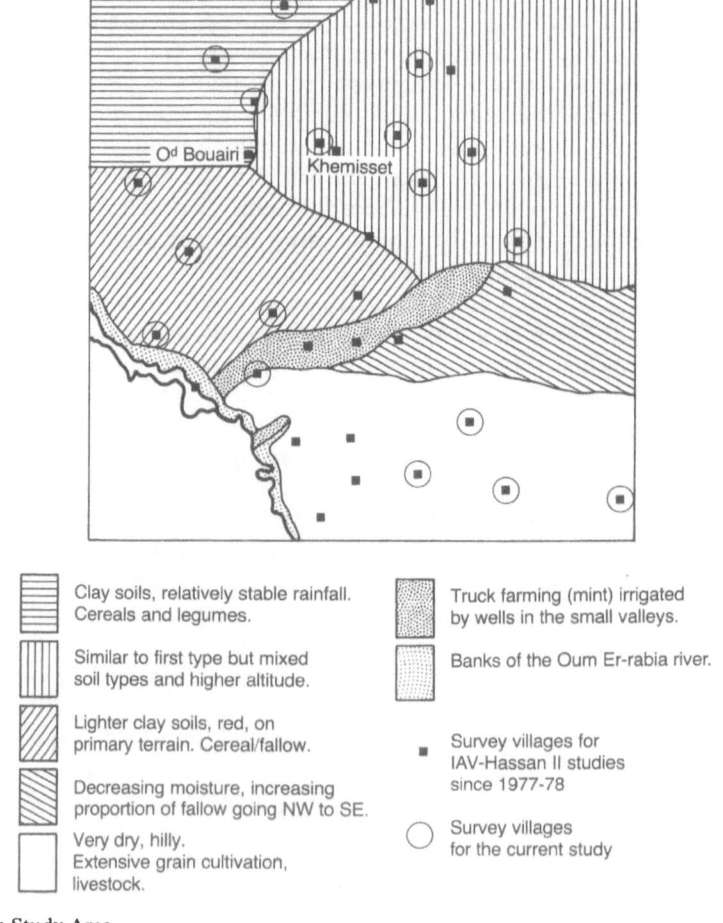

Clay soils, relatively stable rainfall. Cereals and legumes.

Similar to first type but mixed soil types and higher altitude.

Lighter clay soils, red, on primary terrain. Cereal/fallow.

Decreasing moisture, increasing proportion of fallow going NW to SE.

Very dry, hilly. Extensive grain cultivation, livestock.

Truck farming (mint) irrigated by wells in the small valleys.

Banks of the Oum Er-rabia river.

■ Survey villages for IAV-Hassan II studies since 1977-78

○ Survey villages for the current study

Map 1. The Study Area.

rapidly in the area, constitutes one of the most active processes of socioeconomic transformation. In fact, this area was chosen for a vast Program of Applied Research in Dryland Agriculture (Direction du Développement Rural, Institut Agronomique et Veterinaire Hassan II). The farms in the current study are among those followed by the Program from 1981–82 until 1986–87. They had previously been sites for apprenticeships and work-study projects.

The study area extends over 1800 km² and covers a large part of the districts of Settat and of El Brouj (Map 1). There is an increase in aridity from the northern to the southern parts of this area, both with respect to annual rainfall and recorded temperatures. There is also a gradient in soil quality, also running from north to south, with heavy, deep, more productive clay soils in the north. As one goes farther south, the soils are progressively lighter, more gravelly, and less productive. Because of these significant differences in climate and soil conditions, we can divide the study area into two distinct zones, northern and southern.

In the northern zone, favorable conditions result in a greater diversity of farm systems. Autumn grains (wheat and barley) are grown in rotation with maize, legumes, and fallow. A more widespread practice of mechanized agriculture and a greater use of fertilizers indicate a higher level of intensification in this zone, where the yields are in any case greater. The production systems include substantial livestock operations for both cattle and sheep.

The southern zone is less favorable from the point of view of soil and climate. Farming is limited almost exclusively to autumn grain crops that are planted in succession year after year or in rotation with fallow. The rocky nature of the soils and the relatively broken topography of the holdings limits the use of mechanization in this zone. More arid and less predictable, the climate does not permit the use of nitrogen fertilizers that prove very risky for the crops in these conditions. The yields in this zone remain much lower than those in the northern zone. The production systems here are based on the uncertain farming of dryland cereals and the rearing of animals, mostly sheep.

The study sample consists of farms that represent a great variety of climatic conditions, soil types, sizes, and farm orientations with fifteen farms in the northern zone (located in eleven villages) and twenty-five farms in the southern zone (located in eight villages).

Typology of Farm Units

The consequences of agricultural mechanization vary according to the type of farm on which it occurs and the strategies pursued by the farmers. An analysis of these consequences must start with the development of a typology of farms in the two zones. Our typology is based on four criteria that include both quantitative and qualitative variables.

The first criterion is holding size. There are owned holdings that are farmed by the owner, and those that are leased (for a portion of the harvest or a fixed rent) or used in exchange. The holding size was determined for each farm in the sample by calculating the average for the five agricultural seasons from 1981–82 to 1985–86. According to this criterion, five levels of holding size were defined within the sample, from 0–5 hectares to 30 hectares and above.

The second criterion in the typology concerns land ownership. We draw a distinction between farming by landowner and farming by leaseholder to determine how large a portion of the land farmed is under lease. There are two reasons for choosing this criterion. In the first place, one can't help but notice the traditional importance of renting land for the socioeconomic life of arid and semiarid zones. Furthermore, one notes that with the current spread of mechanization, arable lands have become much more desirable. This has resulted in a considerable escalation in land rent which, when paid in kind, can be as high as 50% of the harvest. This criterion therefore allows a distinction between farms whose holdings consist primarily of owned lands and those whose holdings include a sizable fraction of leased lands. In the first case, landowner-farming occurs on at least 70% of the holdings, and in the second leaseholder-farming is found on at least 30% of the holdings. Leased lands constitute on average 30% of the holdings in the study area.

The third criterion of the typology takes into consideration farm ownership of heavy agricultural equipment. Five of the forty farms studied each possess a tractor having as its only accompanying equipment a harrow, a trailer, and sometimes a baler. These farms are among the largest in our sample, with holding size greater than thirty hectares. For farmers without equipment, contract operators perform mechanized operations for a rental fee.

Table 1. Number of Sample Farms by Farm Type and Tenure (Lands Predominantly Owned or Leased).

Type of Farm	Northern Zone			Southern Zone			Total
	Lands Owned	Lands Leased	Total	Lands Owned	Lands Leased	Total	
Micro (1–5 ha)	1		1	3		3	4
Small subsistence (5–10 ha)	1	2	3	5	1	6	9
Medium-sized (10-30 ha)							
In financial difficulty	2		2	3		3	5
In simple reproduction	1		1	3		3	4
Expanding		3	3		4	4	7
Large (30+ ha)							
Unmechanized	1	1	2	2	2	4	6
Mechanized	2		2	2		2	4
Agrobusiness	1		1				1
Total	9	6	15	18	7	25	40

The fourth criterion of the typology is the mode of reproduction of the farms. This criterion serves mainly to distinguish three types of farms (whose holding size is less than thirty hectares) on the basis of their dynamics and evolution: (1) subsistence farms; (2) farms that are simply reproducing themselves; and (3) farms that are expanding.

We found that the distinction between farms of ten to twenty hectares and those of twenty to thirty hectares was negligible. These two levels of holding size were compounded into one: ten to thirty hectares.

Finally, we combined the typology's four criteria to identify for each of the two zones the following farm types (Table 1): (1) micro-farms whose holding size is less than five hectares, mostly farmed by landowners; (2) small subsistence farms (five to ten hectares), with two subtypes: farms where ownership of land held predominates and farms where leasing is a large factor; (3) medium-sized family farms (ten to thirty hectares), with three subtypes: farms in financial difficulty, farms in simple reproduction (in both cases with farming by owner predominant), and expanding farms where leasing is important; (4) large mechanized family farms (more than thirty hectares), with two subtypes: farms in which ownership of land held predominates and farms where leasing is considerable; (5) large mechanized family farms (more than thirty hectares), mostly farmed by landowners; and (6) mechanized agrobusinesses (more than thirty hectares) where ownership of land held is predominant.

Description of Farm Types

At this point we characterize the farm types defined above by describing some of their features, namely family, land tenure (Table 2), and livestock. Our purpose is to highlight the constraints and resources for each of the types of farm.

Microfarms

This farm's basic characteristic is the lack of resources to satisfy the needs of the household and ensure work for the family labor force. Other than the scant holding size (close to 3 hectares on the average, 95% owned by the farmer), this farm on average has only 3 sheep or 0.6 Large Livestock Units (LLU) as income-producing animals, and only one donkey or 0.4 Unit of Labor (UL) as a draft animal. Household size, by contrast, is quite large, both in terms of its aggregate consumption (6.3 Units of Consumption or UC) and its aggregate labor (5.2 Units of Male Labor or UML). Inadequately supplied with private financing, and seldom having the guarantees required by the *Credit Agricole*, farmers are unable to develop a strategy for extending cultivated holdings by leasing additional land. The family labor force must therefore seek outside earnings.

108

Table 2. Area Owned, Farmed, and Leased, by Farm Type and Zone. (Average per Farm from 1981-82 to 1985-86)

Type of Farm[a]	Northern Zone			Southern Zone			Total		
	Owned (ha)	Farmed (ha)	% Leased	Owned (ha)	Farmed (ha)	% Leased	Owned (ha)	Farmed (ha)	% Leased
Micro	3.1	3.1	5	3.3	2.9	5	3.3	2.9	5
Small									
Lands owned	5.2	6.5	15	9.4	8.2	5	8.8	7.9	6
Lands leased	5.0	9.2	53	6.2	9.6	36	5.4	9.4	47
Medium									
In difficulty	19.3	16.1	5	22.8	15.3	5	21.4	15.7	5
Reproducing	19.8	14.1	9	19.3	19.5	6	19.4	18.1	6
Expanding	8.4	20.1	62	14.9	22.1	35	12.1	21.3	46
Large									
Unmech, owned	44.5	51.3	15	41.4	42.2	17	42.4	45.2	16
Unmech, leased	10.8	37.9	71	16.9	39.6	58	14.9	39.0	62
Mechanized	66.6	77.1	20	43.4	43.8	14	54.9	60.4	18
Agrobusiness	57.1	42.8	7	-	-	-	57.1	42.8	7

[a] See Table 1 and text for description of farm types.

Small subsistence farms

The first subtype consists of farms between 5 and 10 hectares, with an average of 8 hectares, 94% of whose holdings are owned by the farmer. These farms have on average 1.2 ULL as income-producing animals (25% of them cattle) and 0.7 UL as draft animals (75% of them donkeys). These farms characteristically have, on average, the smallest households in the sample (4.8 UC and 3.3 UML). Most often, families, whether nuclear or extended, are small but in the process of growing larger. Given the limits of their private resources and the number of family workers, farmers are not able to lease land as a means of meeting their needs. On the contrary, some farmers are unable to cultivate all of their land and must therefore lease some of it out. At the same time, all of these farmers seek outside revenue to support the agricultural enterprise and supply the needs of the family.

The second subtype includes farms with an average size of 9.5 hectares, 47% of which is leased. Each of these farms has an average of 2 ULL as income-producing animals (of which almost 37% are cattle) and 0.8 UL as draft animals (of which 62% are donkeys). These farms also have households whose average size is greater than the households of the preceding type (6.4 UC and 4.7 UML). Having substantial livestock and a larger family labor force, farmers in this subtype often choose to lease additional land to supplement their own holdings, which are insufficient to meet the consumption needs of a large family. But the mobilization of a part of the family to procure external revenue is indispensable for this kind of farm, which is unable to make ends meet only by leasing land: the contracts are very short-term (usually a year) and very high rents are levied (one-third to one-half of the output).

Medium-sized family farms

There are three subtypes in this category, with holding sizes from ten to thirty hectares. Medium-sized farms in financial difficulty are those farms that do not manage to exploit fully and directly their own land resources. The difficulties currently facing these farms result from three major constraints: an insufficiency and/or instability of male family labor, uncertainty about the future because the head of the family is no longer able to manage his farm due to age or state of health, and an inability to obtain sufficient financing. Given these constraints, farmers in this category more often lease out their own land rather than lease land from others. In fact, during the five agricultural seasons (1981–82 to 1985–86), 30% of the owned holdings on these farms were leased out whereas only 5% of the cultivated area consisted of leased lands. With an average ownership of almost twenty-one hectares, this kind of farm finds itself with an average utilized agricultural area of only sixteen hectares.

As for livestock, these farms have an average of 2.3 ULL as income-producing animals (of which 91% are sheep), and 0.6 UL as draft animals (of which 40% are donkeys). The household, even if relatively large (7.5 UC and 5.6

UML), is composed mostly of young people under 14 years of age (50%) and of women (30%), male adults having the smallest proportion (20%). This does not prevent a number of these men from taking off-farm employment.

The second subtype in this category comprises medium-sized family farms that are in simple reproduction. These farms have average owned holdings of 20 hectares, 13% of which are leased out. The average farmed holding is around of 18 hectares, 5% of which are leased. The lands leased out by these farms are often the most remote and/or the least productive. These farms have an average of 3.9 ULL as income-producing animals (of which 25% are cattle), and 1.8 UL as draft animals (of which 45% are donkeys). In comparison with farms in financial difficulty, these farms have more livestock. This increases the use of family labor and at the same time improves the ability of the farm to provide its own financing.

The size of the households on these farms corresponds to the average of the sample for medium-sized farms (8.0 UC and 5.8 UML). Their maintenance at this level is due to the fact that the heads of households are men of advanced years who have extensive traditional know-how and remain quite competent in the conduct and management of their farms. They adopt strategies that tend to promote on-farm work activities and, with the revenue that they generate, satisfy basic household needs. Family members who are employed off the farm most often invest their meager earnings in livestock.

The third subtype in this category consists of medium-sized family farms that are expanding. These farms have an average of twelve hectares of land owned by the farmer, while the actual area farmed is nearly twenty-one hectares. These farms therefore have substantial recourse to leased lands, which represent 46% of their total utilized agricultural area. These farms have an average of 8.8 ULL as income-producing animals (of which 21% are cattle) and, as draft animals, 2.7 UL (of which only 25% are donkeys). Among medium-sized family farms, those in this subtype raise considerably more livestock.

These farms are among the farm types with the largest average size of family units. They have the heaviest burden of consumers (9.2 UC) and the most available family labor (7.1 UML). The owned holdings are often insufficient to provide for the consumption of the household and to make full use of its labor potential. Achieving these two objectives by leasing additional land is a very stable strategy, with many strong points in its favor. In addition to the fact that these farms have a large supply of adult male family labor (40% of total UML), they are managed by enterprising persons who quite often have commercial professions and make very profitable speculations, especially in livestock. These farms also benefit from the support of a number of close family members who live in town and provide the farmers with land and livestock in partnership. Besides being very stable and free of constraint, these partnerships permit the farms to realize substantial revenues. Also, having a financial capacity that is both more diversified and more consistent, these farmers rent land on a fixed-price rather than a share basis. Such lands account for 15% of the total area of lands they lease. Finally, some of these farmers have already managed to accumulate buildings, land, and (through partnership buying) trucks.

Large family farms

There are three types in this category, with holding sizes greater than thirty hectares. The first subtype consists of large unmechanized family farms where farming by the landowner predominates. These are farms whose owned land averages close to forty-two hectares and whose holding size is about forty-five hectares. Of the owned land, 8% is given in exchange as against 2% which is leased out. Of the holdings under cultivation, land taken in exchange accounts for 6%, as against 10% for land which is leased. Farmers make these small readjustments in order to have parcels that are more accessible or more productive, or simply to retain control over portions of land belonging to co-inheritors to whom they pay rent.

For livestock, these farms have an average of 10.7 ULL (of which only 9% are cattle) while draft animals account for only 1.9 UL (of which 38% are donkeys). The average household size of these farms is practically the smallest among all the farm types (6.5 UC and 4.6 UML). Given their resources, these farms easily manage to satisfy the needs of their resident households. Farmers are potentially even in a position to acquire large agricultural machinery, but its management and profitability are problematic.

The second subtype includes large unmechanized family farms with substantial leased lands. These farms have an average of 15 hectares of owned land, and farmed holdings of 39 hectares; leased lands therefore represent 62% of the holdings. The average number of income-generating animals is 11.4 ULL (of which 20% are cattle), while draft animals number 2.8 UL (of which only 17% are donkeys).

The household average size is by a clear margin the largest for all the farm types studied, both as to consumer units (12.2 UC) and adult male workers (8.9 UML). Given the very sizable family burden, farmers opt for leasing additional land. This serves in the first place to satisfy consumption requirements and also to increase the potential use of family labor. Having some of the largest livestock operations, as well as household members whose off-farm activities or professions bring in considerable revenue, these farms have sources of financing that strongly support a strategy of increasing the area under cultivation. Their financing is such that lands rented on a fixed-price basis account for 60% of those in leasehold, and the contracts usually last at least two agricultural seasons.

Some of these farms are equipped with trucks used in commercial and service activities. Some farmers would consider buying a tractor, but only with assistance from the *Credit Agricole*. The size of the farms' land area, however, and the contracts governing the lands under lease (mostly oral contracts, and for a period of less than five years) do not meet the requirements of that institution for granting credit.

The third subtype consists of large mechanized family farms, each of which owns a tractor. The farm households on average have title to 55 hectares of land, of which they lease out 10%; they farm 60 hectares, 18% of which is under lease. Of the land area subject to lease, which includes holdings leased by a farmer and lands he leases out to others, 30% is leased under rental contracts and the remainder under sharecropping contracts.

The rearrangement of land holdings serves a number of purposes and strategies. Farmers lease land in order to increase the potential use of their heavy farm equipment, to improve their production capacity, or to retain control over land belonging to co-inheritors. Lands leased out are usually parcels that are marginal or difficult of access, or else they are designated for maize, a crop which lends itself less readily to mechanized farming and therefore requires more labor than other crops.

For livestock, these mechanized farms have flocks and herds averaging 11.2 ULL (of which 20% are cattle) and draft animals equivalent to 1.7 UL (of which 42% are donkeys).

The households of these farms are among the largest for the farm types identified in the study, both for the number of consumers (10 UC) and the size of the work force (7.3 UML).

In order to buy their farm equipment these farmers have often turned to the *Credit Agricole*, and simultaneously have benefitted from government subsidies. The share of financing supplied by the farm itself comes mainly from the sale of livestock or the harvest, or from earnings sent back by family members working in town. Mechanized equipment performs a large part of the agricultural work on the farm, but also serves as a means of selling services, an important source of revenue on this kind of farm.

Agrobusinesses

In addition to its size, this type of farm is defined by three major features that clearly separate it from all of the types previously described. The owner and his family live in town and not on the farm, a salaried work force is responsible for all farming operations, and the farm is entirely integrated into the market.

In our sample, the only farm of this kind holds title to 57 hectares of land, of which 30% is leased out. Its holdings under cultivation are on the order of 43 hectares, of which 7% is held under lease. Being mechanized and having a family work force only during the months of school vacation, this farm has almost no hoe crops because raising them lends itself poorly to mechanization and therefore requires more labor. Therefore, each year the portions of land designated to these crops are leased out under a sharecropping arrangement. The farmer leases the land because of his concern for protecting the family's patrimony. The land leased belongs to a co-inheritor who lives in town and has no other source of revenue.

For livestock, this farm has 9.6 ULL (31% of which are cattle) and 2.4

UL (17% donkeys). Because the farm is near the urban center of Settat, the farmer has begun to develop a dairy operation. He has drilled a well in order to produce pump-irrigated fodder crops. But the project is still unfinished because of a number of obstacles: a well that is too deep and has inadequate water, a lack of know-how in dairy farming, and the difficulty of maintaining a family and a machine repair shop in town while giving the farm the attention needed to manage such an operation.

The Crop Production System

The employment opportunities offered by the crop production system depend on the crops selected for cultivation and the practices used in raising them. An analysis of employment within the system first requires a description of the crop rotations and an understanding of the uses of mechanization in particular. This approach will make it possible to grasp and analyze other aspects of the farm types already defined.

We selected the two farming seasons of 1981–82 and 1985–86 for our study. They correspond to two quite contrasting situations. The first season was preceded by a severe drought that profoundly disturbed the functioning and equilibrium of the farms. It was less rainy than the second, which ushered in a revitalization of the farms after a period of very unfavorable climatic conditions. Given the situation, an analysis of the farming practices on different kinds of farms, in their respective zones, is relevant for several reasons.

The rotation patterns show that the impact of the 1980–81 drought was much more marked in the southern zone than in the northern, although the latter received less rain in 1981–82 (230 mm as opposed to 310 mm). The drought that preceded the 1981–82 season in the southern zone substantially reduced the land area under cultivation and greatly increased the proportion of fallow. In the northern zone, the only noticeable effect of the drought was a reduction in legumes (planted in winter), which was compensated by an increase in maize planted as a spring crop. Examining the rotations practiced on each type of farm in each zone, one notices a number of differences.

In the northern zone, for the two farming seasons studied (1981–82 and 1985–86), combined autumn cereals (hard wheat, soft wheat, and barley) took up hardly less than 40% of the holdings under cultivation, whatever the farm category (Table 3). The relative portion of each varied widely from one season to the other, but the amount of total field area remained quite stable. We should note that the percentage of field area for autumn cereals is higher particularly on the farm that is an agrobusiness (80% to 100% of the holding), on microfarms (80% to 90%) and on large family farms (50% to 70%). On other kinds of farms, this group of crops takes up only 40% to 50% of the holding.

The place accorded maize introduces another level of difference between the types of farms. On microfarms and on the agrobusiness farm, this crop

Table 3. Crops Grown in the Northern Zone for 1981–82 and 1985–86 (Percentage of Area Farmed).

Type of Farm	Hard Wheat 82	Hard Wheat 86	Soft Wheat 82	Soft Wheat 86	Barley 82	Barley 86	Total 82	Total 86
Micro	–	–	69.4	34.7	19.4	49.2	88.8	83.9
Small
Lands owned	22.6	19.8	15.9	16.5	9.6	12.4	48.1	48.7
Lands leased	10.4	24.3	2.2	11.3	29.2	17.8	41.8	53.4
Medium
In difficulty	9.4	14.3	6.2	23.9	21.8	16.7	37.4	54.9
Reproducing	21.1	15.8	10.6	16.3	8.0	4.2	39.7	36.3
Expanding	17.0	26.4	8.5	5.1	15.6	15.6	41.1	47.1
Large
Unmech, owned	47.6	33.5	–	6.8	13.4	9.8	61.0	50.1
Unmech, leased	10.2	22.1	30.7	3.9	18.4	29.9	59.3	55.9
Mechanized	33.6	28.7	14.8	3.0	22.1	23.0	70.5	54.7
All farms	26.3	25.4	14.4	7.8	18.7	19.3	59.4	52.5
Agrobusiness	18.0	35.8	33.4	27.2	31.8	37.0	83.2	100.0

Type of Farm	Maize 82	Maize 86	Legumes 82	Legumes 86	Herbs 82	Herbs 86	Onion 82	Onion 86	Other Crops 82	Other Crops 86	Forage Crops 82	Forage Crops 86	Fallow 82	Fallow 86
Micro	–	–	3.4	8.0	–	–	–	–	–	–	3.4	–	4.3	8.0
Small
Lands owned	34.4	19.8	–	10.3	1.1	–	10.6	13.9	–	2.9	5.8	4.4	–	–
Lands leased	39.0	19.5	1.4	13.2	–	3.3	–	–	–	–	1.4	4.0	16.4	6.6
Medium
In difficulty	15.6	6.3	–	17.3	–	–	–	–	–	–	3.1	3.9	43.9	17.6
Reproducing	23.8	9.7	–	1.0	–	–	–	–	–	–	–	6.4	36.5	46.6
Expanding	35.4	17.5	1.8	11.6	1.8	0.1	10.3	14.4	–	0.1	1.6	6.1	7.9	3.1
Large
Unmech, owned	24.1	16.7	1.4	12.0	5.4	11.6	–	–	–	–	8.1	9.4	–	–
Unmech, leased	30.7	19.2	6.7	10.0	1.3	–	–	–	–	–	2.0	2.1	–	12.8
Mechanized	14.1	7.2	3.8	16.9	5.7	3.0	–	–	–	–	2.0	5.7	3.1	12.5
All farms	23.2	12.2	3.2	13.6	3.6	2.8	1.6	2.3	–	0.1	2.9	5.5	6.1	11.0
Agrobusiness	3.4	–	–	–	–	–	–	–	–	–	4.6	–	8.8	–

is negligible or absent, while everywhere else it holds second place in field share to autumn cereals. Maize cultivation is greatest on farms with substantial amounts of leased land.

Legumes occupy third place in the rotation pattern. The only significant difference here is that these crops are practically nonexistent on medium-sized family farms that are in simple reproduction and on the commercial farm. In addition, the place taken by fallow is much greater in medium-sized family farms that are in financial difficulty or in simple reproduction.

On the basis of these differences in the crop rotations practiced in the northern zone, we can draw a number of conclusions. On microfarms the holdings are basically reserved for the monoculture of autumn cereals. Lacking sufficient land, farmers seek before all else to satisfy the family needs in regard to cereal consumption. This major constraint leaves only a small share of land for raising other crops. Given also the general inadequacy of resources, farmers opt for the least risky crops. For example, they do not grow hard wheat, a crop that costs more to plant than is justified by the meager yields.

On small subsistence farms, the crop rotations are both the most diversified and the most balanced. They comprise all the autumn cereals in rotation first with maize and next with legumes and/or onions. Unlike microfarms, in which the family work force must provide the outside income necessary for their maintenance, small subsistence farms have crop systems that allow, among other things, greater use of available family labor.

Among the medium-sized family farms, one notices two different situations. Farms that are in financial difficulty or in simple reproduction have relatively large amounts of fallow in the rotation system, which signals the inability of these two kinds of farms to fully exploit their land resources. The size of their herds or flocks cannot alone explain the presence of so much fallow. The second situation is that of expanding farms where one finds crops as diversified as on small subsistence farms.

Among large family farms, the crop rotations show two different situations. On farms where leasing land is important, the rotation pattern is the same as on small family farms and expanding medium-sized family farms (autumn cereals in rotation with maize and legumes). The second situation groups two types of large family farms: those where farming on owned land predominates, and mechanized farms. In these two types, the rotations are marked by the preponderance of hard wheat in the field area given to autumn cereals, by a reduction in maize along with a corresponding increase in legumes, and by the presence of herb crops such as fenugreek and coriander which are destined solely for the market. In this situation, one is witnessing choices that reduce the family character of these farm types. If hard wheat is perceived as a sign of affluence, it is also the cereal whose price at sale is both the highest and the least fluctuating. By reducing maize, which is not readily farmed mechanically and requires more manpower, farmers seek in part to reduce their dependence on hired labor and their recurrent costs. In order to benefit from the appreciable effects of maize as a crop that precedes wheat in rotation, farmers on some mechanized farms and farms with insufficient male family labor annually lease out lands to be planted in maize. The cultivation of crops such as fenugreek and coriander follows a strict market logic. These take up only a very limited space and are farmed sporadically whenever climatic conditions and price are favorable.

The crop rotation on the commercial farm is limited to autumn cereals, particularly wheats. Entirely integrated with the market, this farm specializes in growing cereals, and mechanizes to economize on labor and salaries. Crops such as maize and legumes are raised on lands that the enterprise leases out annually. In this way it receives income, and repossesses these lands the following year to plant them in autumn cereals and so profit from the positive effects of maize and legumes on the soil.

Turning to the southern zone, we find that the crop rotations for all types of farms confirm the primary importance of autumn cereals and fallow (Table 4). The more pronounced aridity of climate in this zone considerably reduces the potential for diversified crops. Barley is generally the most common cereal crop.

A comparison of the two farming seasons shows that effects from the 1980–81 drought were especially felt by small subsistence farms and medium-sized family farms in financial difficulty. On these farms, more than two-thirds of the holding was left fallow in 1981–82. The more favorable climatic conditions of 1985–86 resulted in a general increase in the land area planted at the expense of fallow, which returned to the norm for that zone. These conditions also allowed a slight increase in legumes for nearly all farm types. Increased production of maize occurred only on large family farms.

A closer look at the rotation patterns lets us distinguish three groups of farms: those under thirty hectares with farming on owned land predominant; those over thirty hectares with or without mechanization where farming on owned land predominates; and those of various sizes where leased land is substantial.

The first group seems to raise more barley than hard or soft wheat. Being less secure, farmers choose the crop whose planting costs are the lowest among the autumn cereals, and which is most adapted to the arid climate.

The second group, consisting of large family farms, raises (as in the northern zone) more hard wheat than barley or soft wheat. The sale of hard wheat is more profitable, and its consumption by well-to-do families is higher than for other grains.

The third group of farmers, on the other hand, has a tendency to emphasize barley and soft wheat, even though hard wheat is also cultivated. In making this choice, farmers seem to be balancing three objectives: to minimize risk and to raise cereal crops that are sufficient first in quantity (hence the importance of barley and soft wheat) and next in quality (hence the place reserved for hard wheat). Being highly dependent on leased lands, farmers must first of all reserve a substantial part of their production to pay rent. Having large families, their needs for human consumption are all the more urgent. Mi-

Table 4. Crops Grown in the Southern Zone for 1981–82 and 1985–86 (Percentage of Area Farmed).

Type of Farm	Hard Wheat		Soft Wheat		Barley		Total	
	82	86	82	86	82	86	82	86
Micro	–	14.0	–	–	45.3	42.9	45.3	56.9
Small								
Lands owned	6.5	12.8	2.6	18.4	22.8	35.6	31.9	66.8
Lands leased	16.7	16.8	10.8	34.9	10.0	26.2	37.5	77.9
Medium								
In difficulty	6.5	27.6	–	14.7	15.4	24.7	21.9	67.0
Reproducing	15.2	18.7	2.9	10.2	36.9	40.2	55.0	69.1
Expanding	12.6	17.2	24.3	25.8	30.2	19.9	67.1	62.9
Large								
Unmech, owned	29.4	19.6	10.9	9.5	12.4	18.7	52.7	47.8
Unmech, leased	19.8	9.0	20.8	17.6	27.0	39.9	67.6	66.5
Mechanized	24.9	23.9	4.6	10.2	23.5	31.9	53.0	66.0
All farms	17.7	17.4	10.1	15.6	23.9	29.9	51.7	62.9

Type of Farm	Maize		Legumes		Herbs		Onion		Other Crops		Forage Crops		Fallow	
	82	86	82	86	82	86	82	86	82	86	82	86	82	86
Micro	–	–	–	–	–	0.8	–	–	16.1	3.5	–	11.5	38.6	27.3
Small														
Lands owned	3.0	1.0	–	1.8	1.0	–	–	–	1.0	0.5	–	5.0	63.1	24.9
Lands leased	–	–	–	3.1	–	–	–	–	–	–	–	–	62.5	18.9
Medium														
In difficulty	–	–	–	0.6	–	–	–	–	1.2	2.5	0.6	4.1	76.3	25.8
Reproducing	–	–	–	4.4	–	–	–	0.1	0.8	0.2	–	1.4	44.2	24.8
Expanding	6.4	2.6	–	0.9	–	0.1	–	–	–	1.2	–	4.6	26.4	27.7
Large														
Unmech, owned	–	2.4	1.6	5.3	–	–	–	–	–	–	0.1	6.2	45.6	38.3
Unmech, leased	–	4.5	–	2.2	–	–	–	–	–	–	–	–	32.4	26.8
Mechanized	–	1.4	0.6	2.8	–	–	–	–	–	–	–	1.9	46.4	27.9
All farms	1.1	2.1	0.3	2.7	0.1	–	–	–	0.5	0.5	0.1	3.2	46.2	28.6

nimizing risk and increasing gross production take precedence over the third objective, which is a question of value.

Mechanization in the Choice of Farming Techniques

In the study area, the mechanization process involves four agricultural machines: (1) the tractor and (2) the combine, which came into use towards the end of the 1950s, (3) the maize huller, used since 1970, and (4) the baler, which only came into use in 1975. The pace of progress in mechanization has accelerated since 1970, particularly for the tractor and the combine. However, although the use of tractors has steadily increased, the use of combines and balers is irregular because of crop yields (chiefly of autumn cereals) that fluctuate significantly from one year to the next. The predominance of the disk harrow as the sole implement used in soil preparation and seed covering at the time of planting is another prominent feature of mechanization within the study area.

We should stress that the growth of farm mechanization has created a vast market for the services of private contract operators. In fact, most farmers turn to contractors to perform mechanized operations. As a result, these farmers lose technical control over a large part of the production process. By nature speculative, this form of mechanization has brought few improvements in the quality of work. We are therefore witnessing simple substitution mechanization that is not integrated with a plan for intensification. Although this kind of mechanization cannot produce lasting improvements in land productivity, it has very marked effects on employment. Before opening this subject, let us first examine the sources of energy used on the different types of farm to perform planting, harvesting, and threshing. To do this, we limit ourselves to the most important crops in the rotation (autumn cereals, maize, and legumes) based on the choices made during the two reference seasons, 1981–82 and 1985–86.

In the northern zone, in general, the mechanized planting and harvesting of autumn cereals is fairly standard (Table 5). In the case of maize, the substitution of mechanical means for animal energy and human labor is less common (Table 6). A significant part of the work of preparing the soil and/ or planting this crop still requires draft animals. Maintenance work such as harrowing and hoeing is exclusively performed with the help of draft animals. Although hullers are becoming more generally used, the harvest is entirely performed by hand with a sickle. The cultivation of legumes is the least mechanized, especially for the task of harvesting, which is entirely manual (with or without a sickle), and for threshing, where mechanical means are exceptional (Table 7).

A comparison of the two reference farm seasons shows certain differences in the kinds of energy used. In 1985–86, (1) the use of animal traction in planting maize and legumes has increased; (2) the use of threshers for maize

Table 5. Energy Sources[a] for Planting and Harvesting Autumn Cereals (Percentage of Area Harvested).

Type of Farm	Planting, 1981–82			Planting, 1985–86			Harvest, 1981–82			Harvest, 1985–86		
	M	U	B	M	U	B	M	U	B	M	U	B
Northern Zone												
Micro	100.0	–	–	100.0	–	–	100.0	–	–	90.4	4.6	–
Small												
Lands owned	30.3	69.7	–	100.0	–	–	30.3	69.7	–	100.0	–	–
Lands leased	100.0	–	–	93.4	6.6	–	43.9	56.1	–	79.4	–	20.6
Medium												
In difficulty	100.0	–	–	100.0	–	–	83.3	16.7	–	100.0	–	–
Reproducing	100.0	–	–	100.0	–	–	80.0	20.0	–	75.3	24.7	–
Expanding	88.5	11.5	–	78.2	21.8	–	38.7	61.3	–	78.2	18.5	3.3
Large												
Unmech, owned	100.0	–	–	90.4	9.6	–	96.8	3.2	–	100.0	–	–
Unmech, leased	100.0	–	–	77.1	22.9	–	100.0	–	–	100.0	–	–
Mechanized	100.0	–	–	100.0	–	–	100.0	–	–	100.0	–	–
All farms	97.5	2.5	–	92.4	7.6	–	89.4	10.6	–	95.3	3.6	1.1
Agrobusiness	100.0	–	–	100.0	–	–	100.0	–	–	100.0	–	–
Southern Zone												
Micro	85.8	14.2	–	24.5	75.5	–	–	100.0	–	–	100.0	–
Small												
Lands owned	56.3	43.7	–	43.7	56.3	–	5.4	94.6	–	33.6	42.0	24.4
Lands leased	66.7	33.3	–	–	100.0	–	–	100.0	–	–	100.0	–
Medium												
In difficulty	100.0	–	–	53.8	46.2	–	–	100.0	–	48.1	43.3	8.6
Reproducing	44.9	43.6	11.5	51.2	48.8	–	7.8	92.2	–	52.8	47.2	–
Expanding	55.8	44.2	–	1.6	98.4	–	3.6	98.4	–	20.3	67.0	12.7
Large												
Unmech, owned	78.7	21.3	–	69.5	23.1	7.4	22.0	78.0	–	65.8	34.2	–
Unmech, leased	85.7	14.3	–	93.1	6.9	–	15.4	84.6	–	86.4	10.6	3.0
Mechanized	100.0	–	–	100.0	–	–	28.0	72.0	–	92.7	1.8	5.5
All farms	75.1	23.4	1.5	59.8	39.2	1.0	13.6	86.4	–	59.0	34.9	6.1

[a] M=Mechanized; U=Unmechanized (draft animals or human); B=Both

Table 6. Energy Sources[a] for Planting and Harvesting Maize (Percentage of Area Harvested).

Type of Farm	Planting, 1981–82			Planting, 1985–86			Harvest, 1981–82			Harvest, 1985–86		
	M	U	B	M	U	B	M	U	B	M	U	B
Northern Zone												
Micro	–	–	–	–	–	–	–	–	–	–	–	–
Small												
Lands owned	–	39.2	60.8	–	100.0	–	–	–	100.0	–	–	100.0
Lands leased	100.0	–	–	–	–	100.0	–	–	100.0	–	56.4	43.6
Medium												
In difficulty	100.0	–	–	100.0	–	–	–	100.0	–	–	100.0	–
Reproducing	100.0	–	–	100.0	–	–	–	–	100.0	–	100.0	–
Expanding	–	34.3	65.7	–	16.8	83.2	–	–	100.0	–	72.5	27.5
Large												
Unmech, owned	100.0	–	–	90.0	10.0	–	–	–	100.0	–	–	100.0
Unmech, leased	40.0	–	60.0	–	100.0	–	–	–	100.0	–	–	100.0
Mechanized	100.0	–	–	47.4	–	52.6	–	–	100.0	–	47.4	52.6
All farms	55.7	9.7	34.6	65.3	33.7	1.0	–	3.5	96.5	–	36.7	63.3
Agrobusiness	100.0	–	–	–	–	–	–	100.0	–	–	–	–
Southern Zone												
Micro	–	–	–	–	–	–	–	–	–	–	–	–
Small												
Lands owned	–	100.0	–	–	100.0	–	–	100.0	–	–	100.0	–
Lands leased	–	–	–	–	–	–	–	–	–	–	–	–
Medium												
In difficulty	–	–	–	–	–	–	–	–	–	–	–	–
Reproducing	–	–	–	–	–	–	–	–	–	–	–	–
Expanding	100.0	–	–	–	79.9	20.1	–	100.0	–	–	100.0	–
Large												
Unmech, owned	–	–	–	–	–	–	–	–	–	–	–	–
Unmech, leased	–	–	–	–	–	–	–	–	–	–	–	–
Mechanized	–	–	–	–	–	–	–	–	–	–	–	–
All farms	80.0	20.0	–	–	82.3	17.7	–	100.0	–	–	100.0	–

[a] M=Mechanized; U=Unmechanized (draft animals or human); B=Both

Table 7. Energy Sources[a] for Planting and Harvesting Legumes (Percentage of Area Harvested).

Type of Farm	Planting, 1981–82			Planting, 1985–86			Harvest, 1981–82			Harvest, 1985–86		
	M	U	B	M	U	B	M	U	B	M	U	B
Northern Zone												
Micro	100.0	–	–	100.0	–	–	–	100.0	–	–	100.0	–
Small												
Lands owned	–	–	–	–	–	100.0	–	–	–	–	100.0	–
Lands leased	100.0	–	–	74.3	–	25.7	–	100.0	–	–	100.0	–
Medium												
In difficulty	–	–	–	90.9	9.1	–	–	–	–	–	100.0	–
Reproducing	–	–	–	–	–	–	–	–	–	–	–	–
Expanding	–	86.7	13.3	–	74.7	25.3	–	100.0	–	–	100.0	–
Large												
Unmech, owned	100.0	–	–	79.0	21.0	–	–	100.0	–	–	100.0	–
Unmech, leased	22.6	77.4	–	–	100.0	–	–	100.0	–	–	100.0	–
Mechanized	100.0	–	–	33.4	38.2	28.4	–	100.0	–	–	31.5	68.5
All farms	65.3	33.1	1.0	41.6	40.1	18.3	–	100.0	–	–	69.0	31.0
Agrobusiness	–	–	–	–	–	–	–	–	–	–	–	–
Southern Zone												
Micro	–	–	–	–	–	–	–	–	–	–	–	–
Small												
Lands owned	–	–	–	–	54.5	45.5	–	–	–	–	100.0	–
Lands leased	–	–	–	–	100.0	–	–	–	–	–	100.0	–
Medium												
In difficulty	–	–	–	–	100.0	–	–	–	–	–	100.0	–
Reproducing	–	–	–	–	100.0	–	–	–	–	–	100.0	–
Expanding	–	–	–	–	100.0	–	–	–	–	–	100.0	–
Large												
Unmech, owned	100.0	–	–	–	100.0	–	–	100.0	–	–	100.0	–
Unmech, leased	–	–	–	96.4	–	13.6	–	–	–	–	100.0	–
Mechanized	100.0	–	–	100.0	–	–	–	100.0	–	–	100.0	–
All farms	100.0	–	–	30.7	64.9	4.4	–	100.0	–	–	100.0	–

[a] M=Mechanized; U=Unmechanized (draft animals or human); B=Both

has noticeably diminished; and (3) mechanization in harvesting and/or threshing autumn cereals and legumes has increased somewhat (Tables 5, 6, 7). The first difference is explained by the extreme drought that occurred in 1980-81. Livestock for lease and draft animals were truly decimated by it, so much so that in the following year, 1981-82, the use of machines was for most farmers the only alternative for planting crops. After the calamity, more clement climatic conditions allowed some general rebuilding of herds and a consequent return in 1985-86 to animal traction.

The other two differences can both be explained by the level of yields. In fact, we found that farmers mechanize harvesting and/or threshing less when they estimate that the yields will be too low. In this situation, mechanical means are inappropriate and economically more burdensome. Maize was less productive in 1985-86 than in 1981-82, while the opposite held for autumn cereals and legumes.

The techniques used in planting crops are more apt for bringing out the differences between farm types. They show that, on the whole, animal traction is used on small subsistence farms and on medium-sized and large family farms where leasing land is important. The choice of animal traction results from the special importance given to maize by these different farms. Raising this crop is part of a strategy of putting family labor and team animals to use on the farm as much as possible. Although mechanization plays a large part on these farms in the preparation of soil for maize, only animal traction is used at the time of planting this crop. Manual sowing row by row allows the many maintenance operations (such as hoeing and surface fertilization, which are not mechanized at all) to be more readily performed. This is not the case with the broadcast sowing by hand that occurs everywhere else with the exclusive use of mechanization.

Turning to the southern zone, one should first of all point out that the use of agricultural machinery is on the whole less widespread than in the northern zone and is more common for planting than for harvesting or threshing.

A comparison between the two farming seasons shows that in 1985-86, the use of animal traction for planting crops was greater than in 1981-82 (Tables 5, 6, 7). This change is explained by the rebuilding of animal teams that had been thoroughly decimated by the drought of 1980-81. However, for harvesting and threshing, mechanization (used only for autumn cereals) increased in 1985-86 (Table 5). As in the northern zone, this increase was brought on by an improvement in yields.

Reference to the kinds of energy used for autumn cereals in particular shows that mechanical means are used more on large family farms (those larger than thirty hectares). However, in other types of farms the use of animal energy remains important. This is not the case in the northern zone where nonmechanical farming of autumn cereals is now the exception. The establishment of mechanical means coexists with the cultivation of other crops whose care is more labor intensive and keeps farmers to a tighter work schedule.

The use of farm machinery can be a timely response to events. This was the case during the planting of crops in 1981–82 when draft teams were at first very scarce owing to the extreme drought of the preceding year. Similarly, when projected yields turn out to be satisfactory, farmers decide to mechanize harvesting and threshing.

But stabler aspects of the adoption of mechanization become more apparent with an analysis of the functioning of farms and of the strategies farmers adopt to attain their objectives. The reasons for mechanization mostly depend on the circumstances proper to each type of farm operation. For farms with insufficient team animals and/or available family labor, mechanization is indispensable for maintaining the agricultural activity of the farms. Mechanization can also be a means of freeing the family work force to earn outside revenue, a necessity for the maintenance of the most deprived farms. On some farms where there is insufficient land and a large-scale recourse to leasing lands, mechanization can serve to support this readjustment strategy. The farmed area can increase and, when ecological conditions permit, crop diversity can be greater, which makes possible a more extensive use of family labor. In the case of large farms where wage-earning labor is an important factor, mechanization pays off in the last analysis because it economizes costs.

Our analysis of crop choice and farming techniques brings us finally to examine the implications of these factors for employment opportunities in the crop production system.

Technological Choices and Labor Requirements

In order to quantify actual labor requirements we have used a 'unit of male labor' (UML), which we calculated as follows: 1 UML for men age 15 or over; 0.8 UML for women age 15 or over; and 0.6 UML for children age 8 to 14. These coefficients are applied to each category of worker to obtain the hours of work in UML. By convention, each workday counts as eight hours of work in UML.

In order to understand the relation of mechanization to employment, we must look at farm labor in terms of a sequence of tasks that constitute the ordered and chronological set of all the operations for raising a crop from preparing the soil to shipping the harvest. We will first show the role of mechanization in each operation, then attempt a summary of the whole sequence of work tasks.

The sequence of work tasks can be divided into four main operations: (1) preparation of the soil, e.g., plowing prior to sowing, which can be combined with deep fertilization; (2) planting the crop, which includes applying deep fertilizer, preparing the seed bed, sowing, and covering seed; (3) crop maintenance which includes surface fertilization, weeding, cultivating, and pest control; and (4) harvest (harvesting, threshing, and transportation).

Soil preparation requires only a negligible portion of the total number of

Table 8. Distribution of Labor Time by Farming Period[a], 1981–82.

| Crops | Man Work Days/ha | | | | | | | | | |
| | Northern Zone | | | | | Southern Zone | | | | |
	S	P	M	H	Total	S	P	M	H	Total
Autumn cereals	0.18	0.65	0.32	5.44	6.59	0.13	1.30	2.53	22.65	26.61
Maize	0.07	2.24	7.30	13.56	23.17	1.04	1.28	8.30	9.78	20.40
Winter legumes	–	1.84	0.22	28.19	30.25	–	0.61	–	20.23	20.84
Spring legumes	–	1.26	–	6.41	7.67	–	–	–	–	–
Onions	1.62	24.26	22.79	19.15	67.82	–	–	–	–	–
Other	0.18	1.05	2.22	22.17	25.62	–	5.28	–	6.09	11.37
	Percentage of Each Crop's Total Days									
Autumn cereals	2.7	9.8	4.8	82.7	100	0.5	4.9	9.5	85.1	100
Maize	0.3	9.7	31.5	58.5	100	5.1	6.3	40.7	47.9	100
Winter legumes	–	6.1	0.7	93.2	100	–	2.9	–	97.1	100
Spring legumes	–	16.4	–	83.6	100	–	–	–	–	–
Onions	2.4	35.8	33.6	28.2	100	–	–	–	–	–
Other	0.7	4.1	8.7	86.5	100	–	46.4	–	53.6	100

[a] S=Soil preparation; P=Planting; M=Crop maintenance; H=Harvesting (see text).

workdays (from 0.3% to 5% according to the crop) (Tables 8 & 9). Farmers seldom perform this operation, which is most often done by contract operators; the tractor driver's time is not accounted to the farmer but to the contractor. Therefore, although mechanized plowing has resulted in more land prepared for cultivation, there has been no corresponding increase in employment on the farms.

For all crops, harvesting is the operation that requires the most workdays (between 45% and 85% of total days). The relative share of maintenance operations, on the other hand, depends on the season (from 0% to 40% of total workdays). When the climate is favorable, farmers tend to work more on crop maintenance. In bad years, only maize and onions are properly maintained. In any case, this operation is never mechanized. The different crop maintenance activities call either for manpower alone (weeding, surface fertilization) or for harness animals, especially for cultivating. Planting usually requires less labor than either crop maintenance or harvesting. The workdays given to planting represent between 3% and 16% of total workdays depending on the crop.

The effects of mechanization on employment appear principally in regard to planting and harvesting. These effects (in terms of workdays) depend upon whether these operations are unmechanized, mechanized, or a combination of both (Table 10).

In planting crops, farmers use animal power, tractor with offset disk harrow, or a combination of both. For the study sample as a whole and for all crops,

Table 9. Distribution of Labor Time by Farming Period[a], 1985–86.

| | Man Work Days/ha | | | | | | | | | |
| | Northern Zone | | | | | Southern Zone | | | | |
Crops	S	P	M	H	Total	S	P	M	H	Total
Autumn cereals	0.01	0.60	1.81	1.97	4.39	0.14	1.82	3.06	7.50	12.52
Maize	0.99	3.16	8.50	7.88	20.53	3.55	5.71	8.87	16.64	34.77
Winter legumes	0.30	2.23	4.99	22.77	30.29	0.76	4.13	13.77	32.15	50.81
Spring legumes	0.22	4.24	1.53	11.27	17.26	–	3.62	24.11	10.97	38.70
Onions	0.28	13.84	21.39	13.51	49.02	–	13.00	–	13.00	26.00
Other	0.03	0.36	–	14.93	15.32	–	4.12	–	21.56	25.68
	Percentage of Each Crop's Total Days									
Autumn cereals	0.3	13.6	41.2	44.9	100	1.1	14.5	24.5	59.9	100
Maize	4.8	15.4	41.4	38.4	100	10.2	16.4	25.5	47.9	100
Winter legumes	1.0	7.4	16.5	75.1	100	1.5	8.1	27.1	63.3	100
Spring legumes	1.3	24.6	8.8	65.3	100	–	9.4	62.3	28.3	100
Onions	0.6	28.2	43.6	27.6	100	–	50.0	–	50.0	100
Other	0.2	2.3	–	97.5	100	–	16.1	–	83.9	100

[a] S=Soil preparation; P=Planting; M=Crop maintenance; H=Harvesting (see text)

Table 10. Average Labor Time According to Degree of Mechanization for Planting and Harvesting (Man Work Days per ha).

| | Planting | | | Harvesting | | |
	Non-mechanized	Mecha-nized	Both	Non-mechanized	Mecha-nized	Both
Northern Zone Crops						
Autumn cereals	3	0.5	–	26	2	–
Maize	4	0.3	5	6	–	10
Winter legumes	3	0.5	3	28	–	5
Spring legumes	7	1.0	5	10	–	9
Onions	17	–	20	17	–	–
Other winter crops	2	0.5	2	26	–	14
Other	3	–	3	11	–	–
All crops	5	0.5	7	24	2	10
Southern Zone Crops						
Autumn cereals	4	0.6	2	22	3	10
Maize	5	–	4	13	–	–
Winter legumes	6	0.7	6	26	–	–
Spring legumes	5	0.7	–	11	–	–
Other winter crops	4	–	–	13	–	–
Other	4	–	–	–	–	–
All Crops	4	0.6	3	22	3	14
Crops on all farms	4	0.5	6	22	2	12

the mechanized planting of one hectare takes only a half-day's work (or four hours) as opposed to four days for planting with animal power or with both tractor and animals. The reduction in work time for each hectare planted with a tractor is therefore on the order of 3.5 days. Work times are very nearly the same for both zones in our study area. Paradoxically, there is no noticeable difference between autumn cereals on the one hand and maize and legumes on the other, because of the general practice of broadcast sowing with the latter. In contrast, it takes from twelve to twenty-six workdays for planting onions (seedbed preparation with animal power and transplanting by hand).

The unwillingness of farmers to plant dry (that is, before the first rains), combined with the uncertainty of the climate, means that speed is often the main consideration in planting. The tractor is considered first and foremost for its speed that allows the farmer to put off sowing until there is sufficient rainfall for the crop cycle to get a good start. All these elements push farmers towards an increasing use of mechanization at the time of planting. In consequence, there is a general reduction in employment and more marked seasonal hiring at this time (principally for winter crops).

Farmers use three methods for harvesting their crops: (1) unmechanized: manual harvesting, using animals for transporting and threshing the crop (except in the case of maize, which is threshed manually with sticks); (2) mechanized: using a combine and baler; and (3) a combination of both: manual harvesting and mechanical threshing (with a maize huller, fixed-post thresher, or tractor and harrow).

Mechanized harvesting actually applies only to autumn cereals. For this operation, mechanization effects a drastic reduction in employment: the mechanized harvesting of autumn cereals takes from one to three workdays per hectare as against twenty to thirty days per hectare for unmechanized harvesting. Farmers choose mechanized harvesting of autumn cereals without hesitation when the yields are high. In consequence, the demand for labor is all the greater when the productivity of autumn cereals is low. Moreover, when the yields for autumn cereals are mediocre, harvesting is by hand, usually done by female family members and/or hired labor. The kind of labor used varies according to the means of harvesting. In general, the participation of the female work force increases in direct proportion to the laboriousness of the agricultural work.

Choosing the means of threshing for maize (manual or with the huller) depends in great part on the level of the yields. But in contrast to autumn cereals, the amount of labor required for the harvest increases when threshing is mechanized (from three to nine days per hectare for unmechanized harvesting compared with seven to fourteen days per hectare for combined manual and mechanized harvesting). Maize is the only crop where mechanization does not necessarily lead to a reduction in employment.

Because adequate equipment is not available, crops other than cereals are in general harvested manually. Among these crops, winter legumes (chiefly lentils) require the most labor at the time of harvest (around thirty workdays per hectare).

From these facts, we can assert that the means of harvesting in large part determines the quantity of workdays required for a given crop.

Labor Requirements by Crop

Autumn cereals require an average of 4.4 to 6.6 workdays per hectare in the northern zone and 12.5 to 26.6 workdays per hectare in the southern zone. The difference between the two zones is explained by the greater degree of mechanization in the northern zone. When the sequence of operations for raising cereals is entirely mechanized, cultivation requires three to four workdays per hectare; when the different operations are not mechanized at all, cultivation takes twenty-five to thirty days per hectare.

The amount of labor required to grow maize varies from eleven to twenty-five workdays per hectare, particularly in the northern zone, where most maize cultivation occurs. Machines are used for only two operations: planting (tractor, disk harrow) and threshing (maize huller). None of the other operations are yet mechanized (harrowing, surface fertilization, weeding, cultivating, reaping, postharvest processing). In consequence, the labor needs of maize depend largely on the number of agricultural operations and the level of yields. Maize is a crop with relatively high labor needs, as it requires twenty-one to twenty-three workdays per hectare in a rainy year.

The cultivation of legumes is even less mechanized. Planting is practically the only mechanized operation (tractor, disk harrow); in consequence, mechanization has little impact on the level of labor use. Manual weeding and picking are the tasks that determine the amount of work given these crops. The first task varies in intensity according to the level of weed infestation. But there is an important distinction between winter legumes (lentils, beans, peas) and spring chickpeas. Chickpeas require fifteen to thirty workdays per hectare, but to raise winter legumes takes from twenty to fifty workdays.

Onions require by far the most labor of all crops, from fifty to seventy workdays per hectare. Machine use is generally limited to a tractor and disk harrow for the preparation of the soil. All other operations are performed with the help of a hoe or dibble. We should note that the participation rate of female labor in raising onions is more than 50% of total workdays.

Labor Used in the Crop Production System

In the northern zone, labor requirements for the entire system of crop production are around twelve workdays per hectare regardless of climatic conditions. However, one notices important differences in the distribution of workdays among the various crops raised (Table 11). In 1982, this labor

Table 11. Distribution of Workdays by Crop (in Percentage of Total Man Work Days).

Crops	Northern Zone		Southern Zone	
	1982	1986	1982	1986
Autumn cereals	33.2	22.9	97.6	81.7
Maize	42.2	24.2	1.7	2.8
Winter legumes	7.5	31.1	0.5	12.9
Spring legumes	0.3	5.7	–	0.5
Onions	8.1	10.4	–	0.1
Forage barley	0.6	1.3	0.1	1.8
Other	8.1	4.4	0.1	0.2
Total	100.0	100.0	100.0	100.0

was directed primarily to maize (42%) and to autumn cereals (33%). But in 1986 the same quantity of labor was distributed among winter legumes (31%), maize (24%), and autumn cereals (23%). We can explain the distribution of employment in 1986 by the conjunction of two trends. First, there was a greater use of mechanization in harvesting autumn cereals, which resulted in lower manpower needs for this group of crops. Second, there was a marked increase in the area planted in winter legumes and a corresponding decrease in the area planted in maize, with no notable change in the procedures used. This case shows clearly how the mechanization of cereal production can open the way to greater crop diversity, notably by releasing workers from harvesting.

In the southern zone, the system of crop production is largely dominated by autumn cereals (89% to 97% of the area farmed). The labor needs of the crop production system in this zone are therefore practically identical with those of autumn cereals. Twenty-six workdays per hectare were used in the southern zone in 1982 in contrast to only fourteen workdays per hectare in 1986 over the whole crop production system. This reduction is essentially explained by the greater use of mechanization in harvesting autumn cereals, whose labor requirements for harvesting dropped from twenty-three workdays per hectare in 1982 to eight workdays per hectare in 1986. This decrease in employment affected only hired workers, whose workdays fell almost 50%.

In the southern zone, the topography and the quality of the soils are major obstacles to any significant increase in mechanization. The labor needs of the crop production system in this zone are consequently always higher than in the northern zone, in spite of the latter's greater diversity of crops.

It is certain that if mechanization results in less employment, hired workers are the most affected. But given the fact that not all crop production is mechanized (especially at harvest time), mechanization seems to play a role in regulating the distribution of labor in diversified crop systems.

The Work Calendar

The distribution of workdays throughout the year is determined in large part by the crops selected by farmers and the methods used for cultivation (Figure 1). We have already seen that mechanization causes a sizable reduction in workdays, especially during harvest, and that autumn cereals are the crops for which mechanization is greatest.

There are four periods in the progression of the agricultural year. The first period (September and October) is characterized by an extremely low level of activity in the fields. The workdays in this period account for no more than 2% of the total workdays in a year. The second period includes the months of November, December, and January. It is basically devoted to planting winter crops (cereals, legumes, barley fodder, onions). On average 10% of the workdays for crop production occur during this period. Depending on when the first rains arrive, the high point of sowing is in December or January. The later the rains, the greater the mechanization as farmers use tractors to plant the most acreage in the shortest possible time.

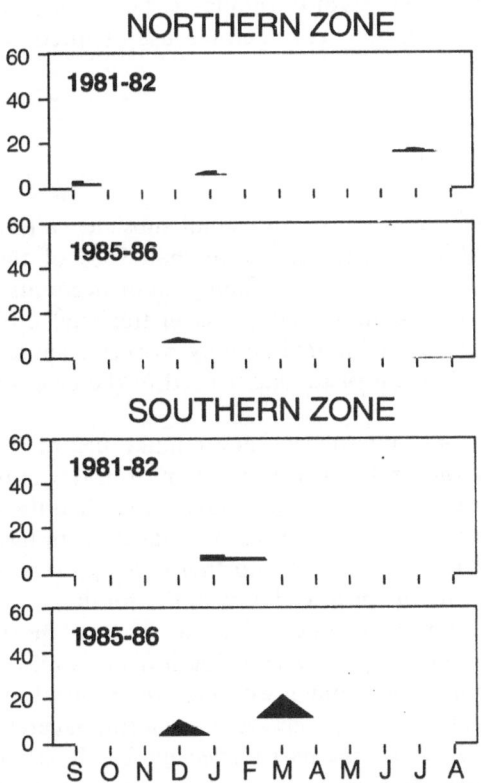

Figure 1. Work Calendar for 1981-82 and 1985-86 Agricultural Seasons (Percentage of Workdays).

The third period lasts from February to April. It begins with the planting of spring crops (maize and chickpeas) in February, the other two months being devoted to the maintenance of all crops. Among the maintenance tasks, manual weeding takes the greatest portion of workdays. On average, 20–30% of the annual workdays occur during this period, with strong fluctuations according to the level of weed infestation. As a general rule, during April the maintenance tasks for winter crops peak along with those for spring crops, which often leads to a substantial use of hired labor.

The fourth and last period (from May to August) is largely given over to harvesting, which occurs in the following chronological order: winter legumes, then autumn cereals and spring legumes, and finally maize. The essential characteristic of this period is that it concentrates at least 50% of the total workdays of the farm year (the rate can reach 85%). This level probably cannot be reduced given the procedures now used. For one thing, the mechanization of harvesting only applies to autumn cereals. Consequently, harvesting the other crops (especially legumes) still requires a large amount of labor. For another, autumn cereals planted on small plots of land are never harvested by machine because of the low level of yields, due either to topography or the rockiness of the holdings. The concentration of workdays during this period results in a considerable use of hired workers and a rise in their daily wages.

Types of Labor

In the northern zone, microfarms and small subsistence farms primarily rely on family labor, which contributes more than 75% of the total workdays (Table 12). For medium-sized farms, family labor accounts for around 60%, with the exception of medium-sized farms in financial difficulty (with only 25%), because of few available male family workers. On large family farms, hired workers provide most of the labor used in the crop production system (60–75%).

In the southern zone, microfarms never make use of hired labor. When farmers need help, they rely on arrangements for the mutual exchange of labor between relatives or neighbors. Small and medium-sized farms first depend on family labor (70–75% of total workdays), with hired workers filling in during the busy periods. As in the northern zone, hired workers constitute the main category of labor for large farms in the south.

Overall, female labor contributes close to 25% of the total days in the crop production system, with the contribution of family members slightly exceeding hired labor. Most women workers are involved in manual weeding and harvesting of cereals and legumes. As most autumn cereals in the southern zone are harvested by hand, the part played by female labor is much greater there than in the north. The employment of women has a tendency to diminish when the agricultural season is relatively good, chiefly because of the widespread

Table 12. Percentage of Labor Used on Sample Farms, 1981–82 and 1985–86.

Type of Farm	Northern Zone						Southern Zone					
	Family		Hired		Exchange		Family		Hired		Exchange	
	82	86	82	86	82	86	82	86	82	86	82	86
Micro	82.0	76.2	5.6	23.4	12.3	0.4	91.9	94.0	0	0	8.1	5.4
Small												
Lands owned	85.5	53.5	14.0	46.4	0.4	0	70.3	79.8	22.2	14.2	7.4	6.0
Lands leased	92.0	87.1	3.0	3.5	5.0	9.3	96.2	59.2	0	40.3	3.7	0.4
Medium												
In difficulty	40.0	19.0	58.9	78.2	1.0	2.7	63.0	78.3	36.0	18.3	0.9	2.8
Reproducing	65.7	50.4	32.0	44.3	2.3	5.2	69.9	73.2	27.8	22.1	2.4	4.6
Expanding	74.0	49.3	21.4	49.5	4.6	1.1	71.5	84.5	27.4	13.5	1.0	1.9
Large												
Unmech, owned	24.4	22.7	75.2	77.2	0.3	0	43.7	35.1	55.6	61.7	0.7	3.1
Unmech, leased	39.9	37.8	60.0	62.2	0	0	53.2	81.8	46.7	16.4	*	1.7
Mechanized	38.2	30.2	55.4	69.8	6.3	0	27.1	54.1	72.8	45.8	*	*
All farms	53.7	39.5	43.4	59.2	2.9	1.1	59.1	71.7	39.2	25.4	1.6	2.9
All farms in sample (north and south)							56.8	57.8	40.9	40.0	2.1	2.1

* <0.05

mechanization of the autumn cereal harvest.

Because of a concern for social status, few women in the households of large family farms work in the fields. Most of the female workers on these farms receive wages. For all other kinds of farms, female family labor plays an important part in farm activity (10–50% of family workdays). On the microfarms, the work of female household members is essential for raising the crops because the males are often occupied in seeking off-farm employment.

Production Costs

In order to discover the effects of mechanization on agricultural revenue, we must first consider the production costs for the main crops and gauge the level of remuneration of family labor.

In the northern zone, the cultivation of autumn cereals is largely mechanized, except in circumstances where the use of animal traction in planting and manual labor for harvesting is dictated by technical constraints rather than economic choices. In this zone, planting and harvesting are completely mechanized for more than 90% of the area planted in wheats and more than 80% of the area planted in barley. A comparison of the costs of production according to manual versus mechanized procedures therefore makes little sense. The costs associated with mechanized cultivation fall into two main areas, planting seed and renting machinery, which together account for more than 90% of the variable costs of autumn cereals. But in contrast to the costs of planting, the costs of mechanization are subject to a rapid rate of increase. Because in this zone farmers most often use their own seed, the main monetary costs of autumn cereals are those of renting agricultural machinery. Nevertheless, the total costs for autumn cereals amount to only 50% of the value of the crop in an average year, and to 35% in years when the climate is favorable. Consequently, the use of agricultural machinery in the northern zone appears to be economical.

In the southern zone, animal power still plays an important part in agricultural operations. In regard to mechanized versus unmechanized cultivation of autumn cereals, there are three main options: (1) planting and harvesting are entirely mechanized; (2) planting is mechanized but harvesting combines manual labor and animal power; and (3) planting and harvesting are performed entirely without agricultural machinery. When both operations are mechanized, the renting of machinery accounts for half of the total costs, and planting for a third. For the other two options, the costs of planting and the wages for hired workers together account for three-quarters of the total costs. When the farmer has no machinery, he primarily relies on family labor, which is free. Thus, total costs are always lower than they are on farms where both planting and harvesting are entirely mechanized. However, in spite of higher costs, this option results in a greater net revenue per hectare because of better yields.

Comparing crop costs in the two zones, one notices that producers in the northern zone have higher costs than those in the southern zone. This is due to four factors: (1) fertilizers are used only in the northern zone; (2) seed quantities used in the northern zone are greater than those used in the south by 30–50%; (3) land rent is higher in the north, around 50% of the crop in contrast to only 33% in the south; and (4) there is greater use of farm machinery in the northern zone, especially for soil preparation and the baling of straw. However, given the difference in the level of yields, farmers in the northern zone always achieve better economic results than farmers in the south.

When we look at the production costs of maize, we find that land rent can account for 40% of total costs, because most maize is grown on land leased by farmers. The cost of wages is next and represents 30–40% of total costs. In contrast to other crops, the cost of planting rarely exceeds 10% of total costs. Because mechanization is secondary in growing maize, the costs of renting farm machinery are never a sizable component of the production costs for this crop.

The cost structure for the production of legumes is similar to that for unmechanized autumn cereals. The cost of planting represents an important part of total costs (up to 50%). However, wages are generally the main expenditure for farmers who grow legumes. For example, the amount paid in wages to harvest a hectare of lentils represents more than twice the total costs for a maize crop. All the same, legumes provide farmers a very comfortable net revenue per hectare, more in any case than autumn cereals or maize.

Revenue from the System of Crop Production

We define family revenue as the difference between the value of all crop production (with the exception of forage crops) and the sum of crop costs (seed, fertilizer, pest control, animal and mechanical power, wages, and rent). Family labor is not remunerated; we therefore consider the net margin calculated in this way as a net profit or revenue earned by the family in the system of crop production. By correlating this revenue with the number of workdays, we have tried to establish an approximation for returns from work by family members.

On average, the workday of a household member in the northern zone is five times more remunerative than in the southern zone: 200 dirhams (MAD) and 40 MAD per day respectively (1 USD = 8 MAD). This disparity is explained by two key factors. First, the net revenue per hectare in the northern zone is twice that obtained in the southern zone. This results from much higher yields and a more diversified crop system in the north. Second, on farms in the northern zone there are half as many workdays by family members per farmed hectare as on farms in the southern zone. This results from the combined effects of a higher rate of mechanization and a greater use of hired workers in the northern zone.

Table 13. Remuneration of Family Labor and Wages for Hired Labor (in MAD).

Type of Farm	Northern Zone			Southern Zone		
	Annual Family Income	Return per Family Workday	Average Daily Wage	Annual Family Income	Return per Family Workday	Average Daily Wage
Micro	5300	163	20	370	8	–
Small						
Lands owned	10500	79	18	2200	29	15
Lands leased	6600	67	22	1100	7	28
Medium						
In difficulty	11300	510	33	2300	24	17
Reproducing	8600	154	15	7300	36	15
Expanding	21600	94	23	5000	21	25
Large						
Unmech, owned	31600	181	23	5700	45	20
Unmech, leased	54800	201	20	21400	83	10
Mechanized	52600	486	27	12000	113	16

In the heart of the northern zone, the family work force is markedly better remunerated than hired workers, whatever the farm type considered (Table 13). But one notices that the level of remuneration for family labor rises as the proportion of hired workers increases. Furthermore, the farm types that have the highest rates of mechanization provide the highest returns for the family member's workday. This is particularly the case for medium-sized farms with few male family workers, and for large mechanized family farms. However, on the strength of revenues from crop production alone, family farms of less than thirty hectares (microfarms, small subsistence farms, and medium-sized farms) would barely be able to meet the consumption expenses of the household. In contrast, large family farms and the agrobusiness achieve sizable surpluses from the crop production system alone.

On the other hand, the various farms in the southern zone generate low revenues. Large family farms are the only ones to obtain a good rate of return on their family labor. Microfarms and small subsistence farms with leased lands are not even capable of remunerating family labor at the market rate. In any case, it is clear that the revenue derived from the crop production system in the southern zone is insufficient to secure the maintenance of the farm, regardless of farm type. This explains the strategic role of animal production and off-farm revenue for the functioning and evolution of the farms in this zone.

Emigration and Off-Farm Work Activities

In this part of the study, we examine the emigration and off-farm work activities of farm family members and their relation to agricultural mechanization. We

Table 14. Rate of Emigration for Adult Males from Sample Farms.

Type of Farm	Northern Zone		Southern Zone	
	Number	Percentage	Number	Percentage
Micro	2	50	4	36
Small	0	0	0	0
Medium				
In difficulty	1	20	6	50
Reproducing	3	60	4	36
Expanding	0	0	2	6
Large				
Unmechanized	4	40	0	0
Mechanized	5	50	7	35
All farms	15	32	23	24

will be concerned only with adult males, who are almost exclusively the individuals who leave to seek new opportunities. For all the farms in our sample, 61% of the adult males have sought sources of alternative income off the farm: 23% have emigrated permanently, 20% are temporary emigrants, and 18% work off the farm without prolonged absences.

One male adult out of four has changed residence and therefore works and lives off the farm. For the sample farms, 65% of these emigrants left between 1974 and 1982. More than two out of five reside in Casablanca, the magnet city *par excellence*, while only 14% of them have settled in Settat, the region's main city, and 5.5% live abroad. Permanent emigration occurs in the northern zone on 43% of the farms and applies to 32% of adult males. In the southern zone, these percentages are 60% and 24% respectively (Table 14). Most of these emigrants come from farms whose lands are predominantly owned by the farmer. They constitute 79% of the emigrants in the northern zone and 95% in the south. Farms with substantial leased lands by contrast have much lower emigration rates, 21% in the north and 5% in the south.

These men make three types of contribution (none of which are mutually exclusive) to their farms of origin: contributions in kind (53% of cases), cash contributions (50% of cases), and contributions in labor (only 8% of cases). These transfers are applied to household consumption in 50% of the cases. They are only very rarely used for the direct financing of mechanized services. A portion of these transfers may be invested in the purchase of animals, and the sale of animal products often contributes to payment for renting farm machinery. But the majority of farmers (90%) do not consider the introduction of mechanization to be the cause of migration. Furthermore, 80% of the farmers we interviewed do not believe that the remittances provided by emigrants have enabled them to mechanize, or that these funds have accelerated the process.

Although temporary or part-time off-farm employment is widespread in arid and semiarid regions, its extent and objectives vary according to the production and employment capacities of available household labor, and to the strategies adopted on different types of farms. Thus, for our sample, we must look at the potential and actual use of male family labor on the farm, by taking into account (1) an estimate of available workdays, excluding days when sick, holidays, visits, market days, rains, and days spent in waiting for the fields to dry; (2) an estimate of days spent in farm work (crops, animal husbandry, and maintenance); (3) an estimate of days expended on off-farm tasks and activities; and (4) a balance of days that are available but not used as workdays, an indication of underemployment (Table 15).

We find that even though the number of male family workers is 50% higher in the south, the level of underemployment is the same in both zones (almost 22%). Furthermore, the average number of days expended on the farm is relatively higher in the north: 273 days compared to 236 days in the south, a difference of nearly 16%. The average number of days employed off-farm

Table 15. Distribution of Male Family Labor in Man Work Days (Average per Farm).

	Days Available	Days Used				Days Not Used	
		On-farm		Off-farm			
		No.	%	No.	%	No.	%
Northern Zone Farms							
Micro	356	45	12.6	–	–	311	87.4[a]
Small	447	219	49.0	190	42.5	38	8.5
Medium							
In difficulty	294	91	30.8	55	18.7	148	50.5[a]
Reproducing	540	347	64.3	90	16.7	103	19.0
Expanding	447	228	51.1	162	36.3	57	12.6
Large							
Unmechanized	820	541	66.0	204	24.9	75	9.1
Mechanized	615	413	67.3	15	2.4	187	30.3
All farms	502	273	54.4	121	24.1	108	21.5
Southern Zone Farms							
Micro	590	108	18.3	472	80.0	10	1.7
Small	543	168	31.0	344	63.3	31	5.7
Medium							
In difficulty	583	198	33.9	322	55.0	65	11.1
Reproducing	727	255	35.0	52	7.2	420	57.8[a]
Expanding	1035	376	36.3	294	28.4	365	35.3
Large							
Unmechanized	1030	312	30.3	610	59.2	108	10.5
Mechanized	960	205	21.4	451	47.0	304	31.6
All farms	766	236	30.8	365	47.6	166	21.6

[a]These high rates result from a defect in the survey.

is, on the other hand, three times greater in the south, 365 days compared to only 121 days in the north.

When we examine the data according to farm type, we find that, as a general rule, the number of days employed in work on the farm increases as the farm's resources increase, as long as crops are diversified and/or the methods used in raising them are less mechanized. In both zones, recourse to off-farm employment occurs in practically all categories of farms. Microfarms (in the south), small subsistence farms, and large unmechanized family farms place the most male family labor in off-farm activities and manage to maximize their employment.

Operating in a climate that is more constraining and permits only rather extensive production systems, and with fewer resources in general, farms in the southern zone have more available male family labor. Being less mechanized, these farms depend upon as much male family labor as the more mechanized farms of the northern zone. But the surplus of workers in this category is still much greater. Also, employment in off-farm activities is considered much more important there (Table 16). In the southern zone, 62% of the household members on 92% of the farms are involved in these activities, while in the northern zone these rates are only 31% and 57% respectively.

However, even though the search for resources other than those generated by farming is a widespread practice, the reasons differ according to the category of farm. On microfarms and small subsistence farms, large-scale recourse to off-farm employment is a response to the inadequacy of the farm's resources in satisfying the needs of the household. These farms cannot maintain themselves without the income from off-farm activities.

On expanding medium-sized family farms and large unmechanized farms, temporary or part-time off-farm employment provides a source of funds for increasing production capacity and wealth. On the other types of farm, outside revenues contribute to the direct needs of the household and farm. But in all cases, off-farm employment is more frequent on farms where most of the

Table 16. Off-Farm Employment for Adult Males from Sample Farms.

Type of Farm	Northern Zone		Southern Zone	
	Number	Percentage	Number	Percentage
Micro	0	0	6	86
Small	3	60	10	91
Medium				
In difficulty	1	25	5	62
Reproducing	1	50	2	29
Expanding	2	40	6	37
Large				
Unmechanized	2	33	11	69
Mechanized	1	20	6	86
All farms	10	31	46	62

lands are leased than in those where farming on owned land predominates: 67% versus 50% in the northern zone and 100% versus 89% in the south.

Off-farm work frequently occurs in towns. In the northern zone, 59% of the men in our sample who also engaged in off-farm activities found work there, while the corresponding percentage in the southern zone was 36%. They are part-time workers (50%), shopkeepers (25%), and traveling merchants (16%). The majority of these individuals find themselves in a variety of situations that fall, in reality, somewhere between being employed and just keeping busy.

Between 1976 and 1985, 68% of all male workers in the sample engaged in off-farm employment. The first half of this period coincided with an advance in agricultural mechanization, while the second was principally a time of drought. Before the introduction of mechanization on their farms of origin, 60% of the male household members had recourse to off-farm employment. But 86% of the farmers polled believed that mechanization reduced the work necessary for cultivating their crops and 68% of them said they had allowed family workers no longer needed on the farm to seek employment elsewhere. Have the revenues derived from off-farm activities made the mechanization of farm operations possible? Only 10% of the farmers answered in the affirmative, while 14% noted that these revenues allowed them a greater use of mechanization. The majority, however, believed that there was no relation between this process and the income obtained from off-farm activities.

In summary, we can conclude that emigration and off-farm employment have occurred on all farm types in the two zones. Both phenomena existed prior to mechanization, although only off-farm employment has increased since then. Although emigration occurs on approximately the same scale in both zones, temporary or part-time off-farm employment is much more widespread in the south where the number of male household workers is higher and the conditions of production more difficult. Farms in both zones rely on an equivalent amount of male household labor for work activities on the farm, although the production systems in the north and south are very different. In addition, the rate of underemployment for these workers is on the order of 20% for each of the two zones.

With respect to farm types, emigration is much greater from farms with mostly owned farmland, whereas workers on farms with substantial leased lands favor temporary or part-time off-farm employment. How the contributions from emigrants or off-farm workers are used depends on the farm's resources and production capacity. On microfarms and on small subsistence farms, remittances are necessary to keep the farms going. This is not the case for medium-sized family farms in financial difficulty or in simple reproduction where these contributions provide a source of savings and investment. But these considerations have not prevented the emergence of a category of young people who attach growing importance to their own autonomy and mobility, who no longer see themselves as being simply members of the collective family work force. These young people seek both economic independence and a different social status.

Opportunities for the employment of male family workers on the farm are greater when a farm's resources are larger and the crops are more diversified. Otherwise, mechanization results in a great reduction in employment opportunities.

Conclusion

Since 1970, agricultural mechanization has tended to spread rapidly in Upper Chaouia. However, it consists of a simple substitution of mechanical energy for human and animal power, and has not brought about an improvement in the quality of work or of the inputs used by farmers. Furthermore, agricultural mechanization is generally in the hands of private heavy machinery contract operators. The producer therefore loses direct technical control over a large part of the production process, being subordinate to the speculative logic of private entrepreneurs. Finally, full-scale mechanization primarily applies to cultivating autumn cereals (hard wheat, soft wheat, barley) and essentially consists of a tractor pulling a disk harrow, a baler, or a combine. All other crops (maize, legumes, onions) are still grown largely with the aid of draft animals and manpower (especially at harvest time). Although farmers may use a rented tractor for planting because of its speed and/or the unavailability of draft animals, the use of combines for harvesting basically depends upon whether yields are high enough to warrant the cost.

From these facts, it is clear that the greatest impact of mechanization on employment will occur for autumn cereals. The mechanized farming of autumn cereals requires only three or four workdays per hectare. The number of workdays per hectare increases to twenty-five when the cultivation of these crops is not mechanized at all. Moreover, we have seen that the greatest reduction in employment comes from harvest mechanization, which results in fewer work opportunities for part-time hired workers. In terms of cost, harvesting by mechanical means reduces by one-half the monetary costs of harvesting autumn cereals.

This said, the adoption of mechanization depends in the last analysis on the conditions prevalent in each of the types of farm. The object of mechanization may be to allow the greater part of the household work force to seek off-farm employment opportunities. Their earnings from these activities are essential for the maintenance of farms whose own resources are insufficient (microfarms and small subsistence farms). Mechanization also allows crops to be raised on those farms that are insufficiently provided with draft animals and/or male family workers (small subsistence farms, medium-sized farms in financial difficulty, and medium-sized farms in simple reproduction). Where ecological conditions permit, mechanization can play an active role in the diversification of the crops, notably by removing seasonal labor constraints, especially at harvest time (the case, in the northern zone, of small subsistence farms, expanding medium-sized farms, and large farms). For large farms in particular, mechanization aims especially at reducing the costs of hired labor, as the farm family members cannot perform all the agricultural tasks themselves.

We can assert that through agricultural mechanization, whatever the purely technical aspects, farmers seek to minimize production costs and/or maximize total revenue, whether it be within or outside of the agricultural sphere. Mechanization results in fewer employment opportunities in agriculture only in regions where ecological conditions prohibit the diversification of crop systems. But, over the middle term, the production systems in arid and semiarid zones are not capable of providing long-lasting solutions to the employment needs of a rapidly growing rural population whose young increasingly desire economic and social independence. There is a risk of seeing the countryside empty of its most active members, those most in a position to contribute to the development of the rural world.

Agricultural Changes on Private Farms of the Sersou, Algeria

AHMED BOUAITA and CLAUDINE CHAULET

Introduction

This case study is part of the research program on the Economics and Sociology of Agriculture and Nutrition. This program of CREAD (Centre de Recherches en Economie Appliquée pour le Développement) is focused on the means of alleviating Algeria's dependence on food imports.

The concerns of the study stem from the research team's previous work (Chaulet 1987), which demonstrated the importance of strategies oriented toward a quest for employment outside agriculture rather than the intensification of production. Now that the job market is shrinking, it is important to know whether farmers have reversed this tendency and started to employ members of their families or hired workers on their holdings, adopting techniques that are more labor-intensive and profitable. It is also necessary to assess whether they are moving in the direction of a recovery of national agriculture and the reduction of underemployment.

Our objective is to understand why farmers adopt any particular productive technique, according to the manner in which they view their interests, the means at their disposal, and the constraints to which they are subjected. Logic determines their strategy, which we will strive to assess by starting with the direct observation of their practices and the relevant explanations they provide.

The case study is therefore based on a direct investigation of the farmers and their holdings, comprising all of the technical and economic elements of the classical agricultural survey. It also focuses on the composition of the family, nonagricultural employment, relations between farmers, and the views of the farmers in regard to their plans.

The region chosen for the study has been dominated ever since the colonial period by extensive rainfed mechanized cereal cultivation. In addition, a certain amount of regional industrialization created extensive nonagricultural employment opportunities, which are now becoming more limited. This past history may shed some light on the relationship between the local labor market and the behavior of farmers.

We deliberately decided not to obtain a representative sample for the region but, on the contrary, to conduct an exhaustive survey of all the farms in two districts selected by means of a 'rational choice.' This methodological

Dennis Tully (ed.), Labor, Employment and Agricultural Development in West Asia and North Africa, 141–164.
© 1990 ICARDA.

option is justified by the poor reliability of the lists of farmers from which we could have drawn a sample. We also wanted to pinpoint the relations and exchanges between the various farms, and to cross-check the statements made by the people surveyed, especially as far as livestock was concerned. This approach was also conducive to the insertion of the investigators into the local society.

An economist, A. Bouaita, and a sociologist, D. Hajj-Ali, conducted the survey in several stages. First, they undertook a preliminary investigation of the area in April and again in July 1987, at the end of an agricultural year characterized by poor rainfall. Second, they administered questionnaires to farmers in November-December 1987, at a time when the autumn rains seemed promising. Data processing charts were compiled from the answers to the questionnaires. The data are currently being analyzed, both in regard to the qualitative aspects of the survey and its economic calculations.

Characteristics of the Region Under Study

The two districts in this case study are Mahdia and Sebaine, both located in the province of Tiaret approximately 300 kilometers southwest of Algiers. These two districts, separated since the administrative reform of 1984, constitute a sociocultural whole that coincides with what is traditionally considered to be the territory of a tribal group, the Beni Lent.

Their total area is about 36,000 hectares; according to the 1987 census the total population of both districts was 30,801 (Table 1). Mahdia, a district center, is an urban agglomeration with a rural hinterland. On the other hand, Sebaine is a rural district whose administrative headquarters is situated in a 'socialist village' (a modern community created during the Agrarian Revolution of the 1970s).

The territory of both districts occupies part of the Sersou, a region specializing in cereals since the colonial period. This region is morphologically part of the High Plateau that elsewhere becomes a desert steppe. However, because of the region's altitude and its exposure to the northwesterly winds, it usually receives adequate rainfall for the cultivation of cereals. The Sersou consists

Table 1. Total Resident Population of the Two Districts Surveyed (1987).

	Resident Population	Number of Employed	% Employed	Number of House- holds	Size of House- holds	No. employed per Household
Mahdia	21,633	3,727	17.2	2,960	7.31	1.26
Sebaine	9,168	1,444	15.7	1,188	7.72	1.22
Total:	30,801	5,171	16.7	4,148	7.42	1.24

Source: ONS (1987).

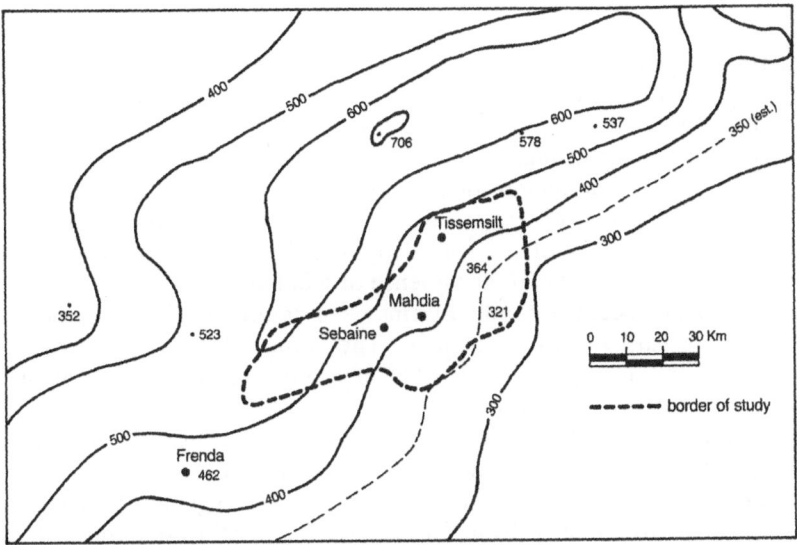

Map 1. The Study Area.

of a range of foothills, the valley of the Nahr Ouassel (the only important river), and the plateau where most of the farmland is situated (Map 1).

Winters are cold and lengthy, with risks of late frost, the summers are very dry, and springtime brings the sirocco. The main constraint, however, is the irregularity of rainfall, particularly the spring rains that have an immediate effect on crop yields.

Before colonization, only the lands on the northern slopes, the banks of the Nahr Ouassel, and favorable low-lying areas were cultivated. The region was mainly occupied by nomadic tribes moving with their flocks and herds according to the season. If we give credence to ancient descriptions, the whole of the Beni Lent population lived in tents. The Sersou experienced belated but intensive agricultural colonization, and the fields of cereal crops obliterated the traditional nomadic routes and expelled the herds.

Beginning in 1880, the earliest colonial speculators started to show interest in the Sersou. Taking advantage of a series of unfavorable harvests, they accumulated large proprietary holdings. By 1900, the Beni Lent had already lost 9,345 hectares; in 1899, a French deputy owned 4,900 hectares of the Sersou. Settlers' villages were founded throughout the region, from 1889 to 1926. (Burdeau, the present-day Mahdia, dates from 1904.) The installation of settlers on medium-sized plots of land rapidly led to numerous failures, but the larger farms had concentrated the land under their control and were able to acquire the means of cultivating it. In 1921, the Beni Lent and their neighbors once again sold thousands of hectares, after the poor crops of the previous year. Twelve 'European' landowners controlled 19,700 hectares in the area.

The region being a harsh one, financial success came only to those settlers who adopted the system of dry farming, alternating the cultivation of cereals with fallow periods, which requires large cultivated areas and extensive mechanization. This mechanization became generalized throughout the region after World War I. The region owes its reputation as a cereal-producing area to the colonial system of dry farming. This system is also responsible for the present poor quality of the soil, deprived of humus by a treatment that exposed it every other year to the violence of wind and rain.

After independence, the colonial farms were nationalized and placed under a system of so-called self-management. They remained large modern farms, characterized by an impressive array of farm machinery and employing few permanent farmhands (about one worker per forty hectares). These 'socialist estates' are currently undergoing reorganization. Both districts (Sebaine and Mahdia) have state-owned and privately-owned land (Table 2).

The history of colonial agriculture in the Sersou explains the constraints and the models to which private farmers were and still are subjected. First of all, the agricultural colonization of the region deprived the local population of the majority of its land. The present meager size of most holdings is a consequence of this. Furthermore, the mechanization of farming on the colonial lands deprived the people of the possibilities of employment. Even the manual labor needed in lentil production seems to have been reserved by priority to groups of nomads, rather than to the local farm families. Furthermore, dry farming reduced the fallow land available as pasture. According to local people, 'the wheat has eaten the sheep.'

Finally, the colonial farms provided a model of mechanized cereal crop cultivation on an extensive scale, a model that many of the colonized adopted. At the end of the 1950s, a geographer pointed out that the number of farms between 100 and 1,000 hectares owned by natives seemed to be increasing. Five of them were from 500 to 1,000 hectares and 33 were between 100 and 500 hectares. All of these farms were held by the Beni Lent and their neighbors, the Beni Maida, in 1951. For the overall region, these large holdings occupied 44.8% of the total area worked by Algerian farmers and 64% of the settlers' lands (Perrin 1960-61).

Table 2. Distribution of Total Utilized Agricultural Area Between Private and Public Agricultural Sectors.

	Number of Hectares		
	State	Private	Total
Sebaine:	20,920	4,710	25,630
Mahdia:	12,624	960	13,584
Total:	33,544	5,670	39,214

Table 3. Number of Unemployed in the Two Districts (1987).

	Casual Laborers		Drivers		Construction Workers		Services		Total:	
Mahdia:	116	(59%)	22	(11%)	43	(22%)	27	(14%)	198	(100%)
Sebaine:	94	(80%)	11	(9%)	13	(11%)	–		118	(100%)
Total:	210		33		56		27		316	

Source: District Administration records (1988).

Current Regional Economic Conditions

There is a network of good roads in the region, and Mahdia is a mere half-hour's drive from the provincial capital, Tiaret. Numerous support services for agriculture are located in Tiaret. These include offices of the provincial administration, grain processing and storage facilities, an agricultural training center, an institute of agronomy, and the provincial office of the agricultural credit bank (BADR). In Mahdia, one can find a branch office of this bank, cooperatives for marketing cereals and legumes, and a flour mill. At Sebaine, there is a silo for the storage of grain. The neighboring district has an agricultural experimental station and numerous agricultural trading firms, in addition to well drillers, cattle dealers, and buyers of garden crops. Thus, all of the necessary conditions for the modernization of agriculture seem to be present even though the operation of this network has yet to become dynamic.

The Sersou has been targeted for industrial development. The city of Tiaret already has some industries (steel, textiles, leather, foodstuffs, building materials) that provide 8,727 jobs. However, most of the jobs are already filled and there are no large industrial sites under construction to generate more. Recourse to nonagricultural employment by farm families, and especially by their young men, is becoming all the more difficult as urban youth are also seeking work. A recent survey conducted by the local administration found 198 unemployed persons in Mahdia district and 118 in Sebaine district (Table 3).

Economic Conditions in Algerian Agriculture

Agriculture in Algeria is characterized by the existence of a dual sector, both at the levels of production and circulation. Land owned by the state amounts to about 3 million hectares of arable land, while that held by private owners constitutes about 4.5 million hectares. Irrigated land is very limited (4% of the total). Sales of land were frozen until quite recently; however, there is an informal market for land (rent, partnership, mortgage, etc.). Joint property held by heirs is quite common. People increasingly rent out stubble fields, fallow land, commons, and grazing rights for cash.

Hired labor is widespread on both state-owned and private farms. On public estates, we find permanent labor and seasonal workers, recruited in principle at peak seasons. The minimum basic wage is 1,300 dinars (DZD) monthly (1 USD = 6.5 DZD), with a share of profits, if any. All hired workers have social insurance and security, and are members of UNPA (Union Nationale des Paysans Algériens). On private farms, there are some forms of 'partnership' that involve sharing the harvest, but most laborers are wage workers who are paid on the basis of piecework or by the day, for short periods. Daily wages can be higher than in the public sector (if the work is demanding), for example, from 100 to 150 DZD in peak seasons. There are also underpaid workers (youth and often women), who receive a daily wage of 50 DZD per day. Workers employed on private farms are very rarely declared to the Social Security Department and are not members of any union.

The state distributes farm machinery and agricultural inputs at fixed and subsidized prices that have increased substantially in recent years. However, as the supply is lower than the demand, a parallel market has arisen, with items available at much higher prices. For example, the official price for a 65-horsepower tractor is 80,000 DZD, but one must pay at least 150,000 DZD to purchase the same tractor on the parallel market.

A dual sector also exists in regard to prices for farm products. The state buys at prices fixed in advance. For example, during the 1987–88 season the prices for durum wheat, soft wheat, and barley were 270, 250, and 230 DZD respectively. We should note that farmers hardly ever deliver barley to government buyers. It finds its way to the parallel market or is used to feed animals.

There is a free market for other products such as garden produce and meat. Prices are established according to forecasts by the buyers, who are often up and about several months before vegetables are ready or lambs have been fattened. During the 1987-88 season, one kilogram of potatoes sold for 4–5 DZD, and one lamb was worth 1,200–1,500 DZD.

Consumer prices are lower than the prices for state-controlled products, of which farmers are increasingly buyers and not sellers. On the other hand, the considerable profit margin of merchants causes prices for consumers of vegetables and meat to be much higher than those received by the producers.

The state agricultural bank (BADR) provides credit at favorable rates (6%). However, many farmers seem to fear getting involved in the credit system, and not all applications are accepted. There are some forms of interpersonal loans, in principle interest-free, but sometimes assuming usurious aspects. It is impossible to obtain accurate data on this point.

This overall economic situation, evoked briefly here to allow comparison between Algeria and other countries, enables us to understand that farmers have two options in regard to increasing their incomes. First, they can place members of their families in nonagricultural permanent jobs. Second, they can produce for the free market (meat and fodder on dry farms, vegetables from irrigated holdings). Within this context of choice, we shall strive to understand the strategy of the farmers of the Sersou as they face new agricultural technologies.

Agricultural Holdings and the Family Labor Force

In the area of the two districts surveyed, we identified 250 privately-owned farms, occupied by families of different structure and size. Our purpose in this section will be to examine the relation between the productive capacities of the holdings and the population that depends upon them, as well as the relation between the labor capacity of the active members of the population and the labor requirements of the farms.

The distribution of land among the farms is unequal (Table 4). The smaller farms represent 58% of the number and 19% of the area (with an average size of 7.5 hectares), whereas the large ones, only 15% of the total number, occupy 50% of the area with an average size of 77 hectares. The remaining 28% of the farms have an average size of 25 hectares.

However, we should note that the farms that we call 'large' are of rather modest size when compared to the former colonial farms or the 'socialist estates' before the agrarian reform of the 1970s, which limited farm size to 32-50 hectares according to ecological zone.

We find irrigated plots on farms of all sizes, although none of the farms are entirely irrigated (Table 4). Thus, we cannot claim that smaller farms use irrigation, whereas the larger size of the others compensates for the low productivity of dry farming. On the contrary, among the largest farms, irrigation is most frequent and requires greater investment.

Under these conditions, the majority of farms, namely the 42% that are smaller than 10 hectares, can hardly feed a family and provide employment for its active members. Conversely, we can predict that the larger farms, comprising both dry and irrigated lands, will have to recruit labor from outside the family.

Although the majority of the farms are owned individually or jointly by families, the largest farms are characterized by a variety of land tenure arrangements (Table 5). One of the largest (110 hectares) is entirely formed of lands that are not individually owned but pooled together from several owners, particularly women. This fact enables us to distinguish the small and medium-sized farms, where most property is inherited, from the large ones, where the surface area varies according to the financial capacity of the farmer.

In this region, as elsewhere in Algeria, there are farm families of different types: nuclear families, joint families of the same generation (two married brothers) or different ones (father and son), or families that combine the latter two types. These forms of social organization, variants of the traditional model of the extended family whose sons bring their wives into the paternal home but whose daughters leave it when they marry, are extremely important for

Table 4. Distribution of Farms by Farm Size Category and Cultivated Area.

Size Category (ha)	Rainfed			Irrigated			Total				
	Number	Total Area	Mean Size	Number	Total Area	Mean Size	Number	% of Farms	Total Area	% of Area	Mean Size
0-5	44	157	3.6	11	42	3.8	55	22	199	4	3.6
5-10	39	308	7.9	11	85	7.7	50	20	393	7	7.9
10-15	27	342	12.7	12	151	12.6	39	16	493	9	12.6
Subtotal	110	807	7.3	34	278	8.2	144	58	1085	19	7.5
15-20	19	342	18.0	10	185	18.5	29	12	526	9	18.1
20-40	26	806	31.0	14	410	29.3	40	16	1215	21	30.4
Subtotal	45	1148	25.5	24	594	24.8	69	28	1742	31	25.2
40+	23	1815	78.9	14	1029	73.5	37	15	2843	50	76.8
Total	178	3770	21.2	72	1900	26.4	250	100	5670	100	22.7

Table 5. Types of Land Tenure on Surveyed Farms (ha).

Size Category (ha)	Individually Owned Area	%	Jointly Owned by Families Area	%	Share-cropped Area	%	Rented or Leased Area	%	Total Area
0-15	796	73	187	17	73	7	5	3	1085
15-40	895	51	607	35	149	9	28	2	1742
40+	1090	38	1390	49	176	6	141	5	2843
Total	2781	49	2184	39	398	7	174	3	5670

the management of income and the mobilization of labor. They also may offer strategies of security or opportunity that affect, among other things, the decisions and activities of farming.

All of these types of families are found on the farms in the two communes, regardless of the size of the farm (Table 6). On the whole, nuclear families are the most numerous, but there are also many joint families of the father-son type. In addition, families tend to be relatively large (Table 7).

The comparison between farms with irrigation and farms whose cultivation is entirely rainfed does not show a tendency for balancing the ratio of population to farm size by the higher capacity of production associated with irrigation. On the contrary, the average number of persons per farm is lower or equal on farms practicing irrigation when compared to rainfed farms of the same size. The number of people and hectares per person is higher only on the medium-sized farms in the irrigated areas. Under such conditions, it is clear that the small and even the medium-sized farms practicing dry farming can hardly satisfy the needs of the members of a family.

In considering the active members of the farm population (Table 8), we follow the definition currently used in Algeria, whereby the active population

Table 6. Types of Family on Surveyed Farms by Farm Size Category (No. of Families).

Size Category (ha)	Type of Family						
	Nuclear	Nuclear With One Relative	Joint (father-son)	Joint (brothers)	Complex	Other	Total
0–15	63	22	44	7	8	0	144
%	44	15	30	5	6	0	100
15–40	19	9	24	6	1	10	69
%	28	13	35	9	1	14	100
40+	14	9	7	4	3	0	37
%	38	24	19	11	8	0	100
Total	96	40	75	17	12	10	250
%	38	16	30	7	5	4	100

Table 7. Farm Population by Farm Size Category.

	Size Category (ha)			
	0–15	15–40	40+	Total
Number of farms	144	69	37	250
Total area (ha)	1085	1742	2843	5670
Total population	1381	772	372	2525
Total population (%)	55	31	15	100
Persons/farm (mean)	9.6	11.2	10.1	10.1
Mean ha/person	0.8	2.3	7.4	2.25
Rainfed				
Persons/farm (mean)	9.3	11.7	10.0	
Mean ha/person	0.8	2.2	7.8	
With Irrigation				
Persons/farm (mean)	8.0	10.0	10.0	
Mean ha/person	0.8	2.4	7.3	

Table 8. Active and Employed Members of Farm Population by Farm Size Category.

	Size Category (ha)			
	0–15	15–40	40+	Total
Number of farms	144	69	37	250
Average size (ha)	7.5	25.2	76.8	22.7
Active population	335	185	93	613
Number employed				
Total	278	160	76	514
Per family	1.9	2.3	2.1	2.1
Number unemployed				
Total	57	25	17	99
Per family	0.4	0.4	0.5	0.4
Percentage of active employed	83	86	82	84

includes able-bodied men between the ages of 18 and 60, and only those women who receive remuneration for specific activities. There are roughly two employed persons per family on the farms in the survey. The work load per employed person is highest for the families living on the small farms. This heavy burden is common in Algeria, taking into account the number of children per couple and the convention that excludes women from the active population.

Farmers are frequently engaged in off-farm employment, regardless of the size of the farm (Table 9). In fact, off-farm employment represents about one third of the total employment, with one job per family. This fact is important, for although it enables us to understand how families on small farms can subsist, it also shows that even for the large farms, we cannot

Table 9. On- and Off-farm Employment of Active Farm Population by Farm Size Category[a].

| | Size Category (ha) | | | |
	0–15	15–40	40+	Total
Active population	335	185	93	613
Total number employed	278	160	76	514
Employed on farm				
Full-time	129	86	52	267
Part-time	68	34	10	112
Ha/employed	5.5	14.5	45.6	15.0
Employed off farm				
Permanent	88	52	16	156
Seasonal	45	15	2	62
Farm work	31	14	2	47
Construction	48	14	5	67
Industry	13	11	0	24
Services	41	28	11	80
Percentage employed				
off farm	48	42	35	42

[a] An individual may work part-time on the farm and also work part-time or full-time off the farm.

understand agricultural behavior unless we take off-farm activity into account in terms of working capacity and investment.

On the other hand, we see that the quality of employment is linked to the size of the farm. For example, part-time jobs, whether on the farm or outside, are more frequent for the smaller farms. Furthermore, wage labor on other farms and public works (pick-and-shovel jobs) are the lot of the small farmers, whereas the families of the large farms send their members to work in the civil service. Industrial jobs are relatively important for the families of the medium-sized farms. Therefore, the opportunity for access to external employment is unequally distributed among the families, according to the size of their holdings. Thus, the fruits of off-farm employment cannot fully compensate families for the unequal distribution of land.

The presence of irrigated land is associated with greater employment on the farms (Table 10). The number of hectares per employed family member on the farm is always lower when irrigation exists, very clearly so (from sixteen to nine) for the medium-sized farms. On the other hand, the burden of temporary employment diminishes on all farms with irrigation. The fact of having irrigated land is thus a necessary condition for the improvement of the internal rate of employment among the members of the family, although this improvement may not be markedly felt because, on the one hand, only part of the holding is irrigated and, on the other, the increasing need to cultivate irrigated land can be met by hired labor from outside the farm.

One of the preliminary questions that we can raise in regard to farm family labor is its relationship to mechanization. We find that all owners of the

Table 10. On-Farm Employment of Active Farm Population on Rainfed or Irrigated Holdings.

| | Size Category (ha) | | | | | |
| | 0-15 | | 15-40 | | 40+ | |
	Rainfed	Irrigated	Rainfed	Irrigated	Rainfed	Irrigated
Number of farms	110	34	45	24	23	14
Total area (ha)	807	278	1147	594	1814	1028
Full-time employed	86	43	52	34	30	22
Part-time employed	51	17	22	12	7	3
Total employed	137	60	74	46	37	25
No. employed on-farm per farm	1.2	1.8	1.6	1.9	1.6	1.8
Area/employed (ha)	5.9	4.6	15.5	12.9	49.0	41.1

Table 11. Use of Mechanized Farm Equipment on Surveyed Farms by Farm Size Category.

| | Size Category (ha) | | | | | | | |
| | 0-15 | | 15-40 | | 40+ | | Total | |
Number of Farms	Rainfed	Irrigated	Rainfed	Irrigated	Rainfed	Irrigated	Rainfed	Irrigated
Do not use equipment	0	0	0	0	0	0	0	0
Use rented equipment	108	31	32	20	5	2	145	53
Use own equipment	2	3	12	5	18	12	32	20
Total	110	34	44	25	23	14	177	73

private farms in the districts surveyed use farm machinery in their operations, at least for tilling and harvesting. They also use hired combines for the accessible plots of land.

For the majority of farms (and for nearly all of the smaller ones), farmers rent machinery from private firms, other farmers, or private contractors who have acquired machinery sold off by parastatal agencies that have been dissolved during the past few years. The majority of the large farms have their own equipment (Table 11).

This reliance on mechanization for the dominant dry-farming cultivation system (cereals-fallow pasture) reduces the necessary work per hectare, which is often reserved for hired laborers. Thus, when farmers rent tractors, they also hire people to drive them. On the small and medium-sized rainfed farms, the farm work by members of the family is thus reduced to a few days per year; on the larger dry-farming units, there is virtually no family labor at all.

Such a situation was acceptable, even desirable, in former times when families could purchase their food cheaply, when men could find outside employment under satisfactory conditions, and when farmers could rent machinery at favorable rates from service cooperatives or the 'socialist estates.' At present, however, the relative equilibrium obtained by farm families who combined agricultural activities with off-farm employment has been disrupted. Having lost all hope of obtaining employment for the young men of the family outside the farm, and having to assume the burden of higher expenses for yields that remain low, will farmers intensify their agricultural production in order to provide on-farm employment for the members of their own families and increase their yields?

Under conditions prevailing in the region, and with an extensive mechanized system of farming, farmers can try to increase on-farm employment for their families and boost family income by adopting one of the following strategies: (1) the intensification of cereals cultivation, with a crop rotation system of cereals-legumes and/or fodder; (2) the introduction or intensification of animal husbandry, pursued along with cereals cultivation; (3) drilling new wells and irrigating all or part of their acreage. All of these strategies require investment and a change of attitude toward farm work on the part of family members. Moreover, as a prerequisite, there must be favorable market conditions and prices.

An examination of the results of our survey will enable us to ascertain the behaviors of the heads of the farm families in regard to these strategies. We can then pinpoint the constraints that have prevented the necessary changes from occurring.

Intensification of Cereals Cultivation

In striving to create on-farm employment opportunities for the active members of his family and to increase the family's overall income, the first option for a farmer in a region noted for the production of cereals would be to intensify cereal crop cultivation. The means of intensification form a sort of 'package' of interrelated practices that include improved methods of soil preparation, the use of improved seed, the application of chemical fertilizer, and mechanized harvesting. This improved form of cereals cultivation should occur within a pattern of crop rotation where fallow pastures are replaced by the cultivation of crops (legumes or fodder). However, the intensification of cereals cultivation is not very common in this region.

We have identified three levels of intensification in the region: (1) Level C, where farmers do not apply fertilizer, and use their own seed or seed purchased from other farms; (2) Level B, where farmers make an effort to intensify their cereal crop cultivation with better preparation of the seedbed and a frequent use of fertilizer; and (3) Level A, which approximates the optimal utilization of all recommended inputs.

In examining the breakdown of surveyed farms according to level of intensification and size of farm, we observe that 4 out of 247 farms are using cultivation techniques close to the recommended package, while 33 farmers are implementing part of the plan. However, the intensification of cereals cultivation occurs primarily on large farms. The 37 farms using all or part of the recommended package are distributed as follows: 7 out of 141 farms smaller than 15 hectares, 12 out of 69 farms between 15 and 40 hectares, and 18 out of 37 farms larger than 40 hectares.

The majority of the holdings are extensive, including half of the large farms. However, most of these farms have agricultural machinery. Consequently, these farmers are in a position, in principle, to apply the recommended techniques without excessive costs. The possession of the equipment is a necessary condition, yet not sufficient for the intensification of cereals cultivation. In addition, farmers make very limited use of fertilizer and do not use potentially high-yielding varieties of wheat or barley.

The other alternative for the intensification of dry farming would be the introduction of a crop rotation pattern that includes fodder and legumes. However, this practice is nearly absent as well, regardless of the size of the farm (Table 12). In spite of efforts to introduce fodder on the state farms,

Table 12. Crops Cultivated on Surveyed Farms by Farm Size Category.

Size Category (ha)	Oats (ha)	Legumes (ha)	Total Oats & Legumes (ha)	Wheat & Barley (ha)	Oats & Legumes/ Wheat & Barley (%)
0-15	66	0	66	544	12
15–40	110	8	118	802	15
40+	211	0	211	1434	15
Total	387	8	395	2780	14

and in spite of the memory of the lentils produced by the settlers of the Sersou, the farmers remain attached to their current rotation pattern of cereals-fallow pasture. They explain that vetch does not grow well and that the value of legume harvests does not justify the labor costs involved. We should note that the only farmers who grow legumes have medium-sized holdings and make the greatest use of the family labor force.

We can thus conclude that it is not by intensifying their cereals cultivation or changing their dry-farming crop rotation patterns that the private farmers of the region are striving to increase family employment opportunities or family incomes.

Livestock

Animal husbandry is an important enterprise for most of the farms in our survey (Table 13). Only 74 out of 250 farms do not raise any livestock other than a few sheep and/or one or two cows.

Raising sheep is a traditional enterprise in this region and, under present market conditions, is much more profitable than growing cereal crops for sale at official prices. We found that 170 of the surveyed farms (69%) have substantial numbers of sheep (Table 14). We did not take into account households with one to four sheep; while an important family resource, they are not part of an agricultural enterprise in the strict sense. The farms are distributed roughly equally according to size. However, the average size of the flock is greater on the larger farms. In addition, the difference between the average flocks of small farms and those of large farms is less prominent than the difference in farm sizes. This observation leads us to ask whether the presence of a flock is likely to compensate for the small size of a holding, both in terms of income and employment.

Even though the size of the flock corresponds roughly to size of farm (Table 15), flocks of average size (21 to 60 sheep) are present on 28% of the small farms and on 45% of the medium-sized ones, for which they certainly constitute a financial contribution that is relatively substantial. In addition, four small farms and seven medium-sized ones have relatively large flocks, a fact that

Table 13. Animal Husbandry on Surveyed Farms by Farm Size Category.

Size Category (ha)	Number of Farms	Farms with No Livestock		Farms with Sheep	Farms with Cattle	Farms with Poultry
		No.	%			
0-15	144	44	31	100	70	3
15-40	69	19	28	44	43	1
40+	37	11	30	26	19	4
Total	250	74	30	170	132	8

Table 14. Raising Sheep on Surveyed Farms by Farm Size Category.

Size Category (ha)	Number of Farms	Number of Sheep	Farms with sheep		Mean Size of Flock	% of Farms With Sheep
			No.	%		
0-15	141	2153	100	59	22	71
15-40	69	1705	44	26	39	64
40+	37	1734	26	15	67	70
Total	247[a]	5592	170	100	33	69

[a] Does not include three farms that specialize in poultry.

Table 15. Distribution of Flock Size by Farm Size Category.

Number of Sheep	Size Category (ha)							
	0–15		15–40		40+		Total	
	Number	%	Number	%	Number	%	Number	%
5–10	37	37	6	14	0	0	43	25
11–20	31	31	11	25	2	8	44	26
21–35	16	16	11	25	8	31	35	21
36–60	12	12	9	20	8	31	29	17
61–99	3	3	4	9	5	19	12	7
100–250	1	1	3	7	3	12	7	4
Total	100	100	44	100	26	100	170	100

enables us to consider them as farms specializing in raising sheep.

Next, we need to ask how these sheep are fed. When we calculate the number of sheep per hectare for these farms, we find that the ratio of sheep per hectare is relatively low (0.8) for farms larger than 40 hectares. These farms may have enough fallow, stubble, straw, and barley to feed their sheep. On the other hand, this ratio is very high (3.0) for farms smaller than 15 hectares. Under such conditions, farmers can feed their flocks only by relying on external resources (collecting straw from the fields of state farms, pasturing their animals there or on rented pasture, buying animal feed). On these farms, raising sheep appears not as the complement of agricultural activity, but as an enterprise conducted with its own logic and within its own system of relationships and specific exchanges.

In the area of our survey, cattle are not as widely raised as sheep; 132 farms have herds of different sizes. Altogether, there are 384 head of cattle (Table 16). The majority of farms with cattle (53%) are those smaller than 15 hectares. However, the number of cattle per farm is generally quite low. Only 15 out of 60 small farms have herds of more than 3 head, i.e., 21%

Table 16. Raising Cattle on Surveyed Farms by Farm Size Category.

Size Category (ha)	Cattle		Farms with cattle		Mean Size of Herd	% of Farms with Cattle
	No.	%	No.	%		
0–15	156	41	70	53	2.2	49
15–40	133	35	43	33	3.1	62
40+	95	25	19	14	5.0	51
Total	384	100	132	100	2.9	53

Table 17. Number of Farms Raising Poultry by Farm Size Category.

Size Category (ha)	Poultry for Meat	Poultry for Eggs	Total
0–15	2	1	3
15–40	0	1	1
40+	4	0	4
Total	6	2	8

of the herds that can be considered as having a market value. On farms between 15 and 40 hectares in size, 49% of the herds are composed of more than 3 head of cattle. Finally, for the largest farms (forty hectares and over), 74% of the herds have over 3 head. Furthermore, three of these farms have herds that include over ten cattle of various imported breeds.

A few farmers in the surveyed area raise poultry on a commercial basis (Table 17). Traditionally, women were in charge of this activity, which provided a source of income and a certain degree of economic independence. At present, family poultry farming continues to thrive, but the enterprises are larger, more of the final products (eggs, meat, chicks) are sold on the market, and men have a greater role in management.

The presence of livestock on the farm raises the problem of feeding it. Farmers can purchase feed on the market, but at high prices they can ill afford. The needs of the herd or flock can therefore intervene to determine choices in regard to agriculture and techniques of cultivation.

The dominant agricultural model in the area is essentially based on a combination of cereal crop cultivation and raising sheep. The diversification of agricultural activities on the farms, although not negligible, remains somewhat limited. Raising sheep, however, remains the dominant farming activity in the region. This form of animal husbandry, as long as it is extensive, offers very little scope for employment. Over 80% of the flocks on the farms surveyed have fewer than 40 sheep, and thus do not provide a full-time occupation for an active worker. On the other hand, looking after these sheep can be an opportunity for members of the family who cannot find employment elsewhere (e.g., elders or children who have left school). Sheep are profitable only in the sense that they enhance the value of the by-products of cereals cultivation and the fodder from pasture, obtained at low cost and often at no cost at all.

Nevertheless, on the small farms, the size of the flocks, although not large, exceeds the productive capacity of the farm in providing the necessary animal feed for their daily needs. In order to meet these requirements, farmers leave part of their lands fallow throughout the year, an area that may represent as much as 50% or more of the arable land.

Other factors explain the maintenance of uncultivated fallow land. Working fallow land is very costly, especially as the effects of this practice on crop yields and incomes can be insignificant. Renting a tractor to work a hectare normally left fallow means an increase of 36% in the expenses per hectare, whereas the corresponding increase in yield is uncertain. Economic logic is at the root of the importance of raising cattle or sheep and maintaining extensive cereals cultivation.

The prevailing agricultural practice of cereals cultivation and raising sheep is characterized by (1) a certain proportion of farmland left fallow and (2) a certain proportion of cereals acreage whose product becomes food for animals.

In regard to the first characteristic, we can distinguish three categories of farms in our survey according to percentage of total utilized agricultural area left fallow. The first category consists of 30 small farms and 4 large farms where the percentage of fallow is below 40% on each farm. The second category includes those farms where the percentage ranges from 40% to 50%. The majority of the farms (181) fall into this category, and 53% of them are small farms. The third category comprises 35 farms with over 50% of their total utilized agricultural area left fallow. Twenty of these farms are medium-sized (15-40 hectares) and the rest are smaller than 15 hectares. In general, we note that the amount of land left fallow does not increase with the size of the farm.

In regard to the amount of cereals acreage destined to produce animal feed, we find that the majority of the farms (144 or 58%), including all of the small farms, have over 60% of the total area for cereals cultivation set aside for this purpose. In addition, 28 medium-sized farms (15–40 hectares) have designated 50–60% of their cropland for feeding animals. Finally, 78 farms, both medium-sized (59) and large (19), use only 30–50% of their cultivated area to sustain their flocks and/or herds. To summarize, we may say that 70% of the farms surveyed set aside a large portion of their land (70–80%) to meet the needs of livestock.

In the region of our survey, it seems that after a period of specialization in cereals cultivation introduced by colonization, farmers have returned to their erstwhile specialization in livestock raising, with some farmland and crop yields used as a means to feed animals. This response is based on a rational calculation by farmers according to the present state of the market and of agricultural technologies, which are still incapable of offering reliable technical solutions to the problems of farmers in semiarid zones with an irregular climate. However, this response is not a solution to the problems of employment. In fact, it offers only complementary activities to farmers who are highly underemployed.

Market Gardening

In the early 1980s, the government distributed motorized pumps and water pipes free of charge to fifteen farmers who were starting to cultivate irrigated garden crops for the market. Between 1982 and 1987, fifty farms in the area received bank credits for sinking wells and purchasing irrigation materials. In addition, some merchants financed irrigation in partnership with farmers. However, the practice of irrigated agriculture remains low, if one takes into account the number of farms with irrigation, and especially the size of the irrigated plots in relation to the total utilized agricultural area of the region. The adoption of irrigated agriculture is blocked by the irregular availability of water, difficulties in obtaining seeds, and problems in marketing the food products. These problems are aggravated by the fact that farmers have a tendency to concentrate on a single product, potatoes.

If we admit that irrigated farming could be the point of departure for important transformations, we can also say that specialization in market gardening, especially monoculture, is not justified for a number of reasons. First, farmers become overly dependent on a market that is often saturated. Second, irrigation requirements are considerable, whereas water is often scarce at periods when the heat is very intense. Finally, the shallow soils do not permit high yields. Although the added value per hectare is high, in relation to cereals cultivation, the yields per hectare remain low and the income for a day's work does not exceed 70 DZD. (One hectare of market gardening occupies one person for 90 days, which gives an average income per hectare of 5,600 DZD.)

In addition, we should note that the introduction of new market crops occurs within the framework of complex social relations that are not always favorable to the integration of market gardening into the overall operations of the farm. To a great extent, the producer is subject to the whims of the merchants who buy his crops. These can intervene as lenders to launch the campaign and, in a way, control production. The producer may also enter into partnership with a landowner, each sharing half of the proceeds. Conversely, he may lease a plot of land for the purpose of practicing market gardening. In addition, instead of using family labor, market gardeners frequently resort to hired laborers paid on a piecework basis, often young unemployed persons, who complete the work quickly in order to meet the demands of the cultivation schedule and the requirements of the market. Thus, one can find families whose members are underemployed having recourse to hired labor.

Such a form of introducing new agricultural activities seems to correspond more to an immediate quest for additional income than to the constitution of a well-balanced farm. The clearest sign of this is the focus on potato monoculture, an agronomic heresy illustrated by the yearly succession of two crops on the same plot without fertilizer or herbicide. For example, farmers who alternate potatoes with tomatoes or peppercorns are rare. During the survey, we saw only one farmer cultivating fodder legumes in rotation with other crops on irrigated land.

The contrast between the substantial investment required by irrigation and the perfunctory agronomic practices that follow it deserves some reflection. It is not a question of ignorance (although the official agricultural services could be more active); farmers conduct the actual irrigation operations with care. It appears instead that farmers, because of the constraints of soil and water, are not in a position to make irrigation a crucial factor for restructuring the operations of their farms.

The Dynamics of Farming

We noted earlier that the privately-owned farms of the Sersou did not adopt the package of cereal crop intensification. In spite of this, they are neither 'traditional' nor stagnant. Raising livestock with the use of produced or purchased animal feed, and irrigated market gardening, where this is possible, are options of intensification that ensure increased income and allow the farmer to reduce the underemployment of his family labor force.

We have also noted that these options are limited by external constraints to which farmers are subjected (irregular climate, soil conditions, supply of water, availability of means of production, lack of organized markets). Because of these constraints, the results obtained are far from adequate for solving the problems of local employment and increased agricultural production.

What factors are likely to support the farmers' efforts and enable them to develop their farms?

Certain interventions in the region by the government could have a positive impact on the well-being of farmers. For example, the government has made an agricultural investment with the construction of the Dahmouni Dam on the Nahr Ouassel (Sebaine district). This dam will have the capacity to retain 30 million cubic meters of water, and should enable the population to irrigate 3,000 hectares of land. However, it is not possible to ascertain exactly what its effects on production and employment will be (although seventy local youths are working on construction at the site). However, there are estimates that the dam, when completed, will double the present rate of agricultural employment.

On the other hand, the current reorganization of the public agricultural sector should result in more land available to farmers, on the basis of approximately forty hectares per farmer (recall that we considered as 'large' those private farms exceeding forty hectares). Because of the lack of precise data, we cannot elaborate on this point. It is likely that the beneficiaries of this land reform will raise livestock, which may translate into fewer food resources for neighboring farmers who formerly used the products of the state-owned farms. Finally, the industrial development planned for the region is likely to create new opportunities for employment and public works jobs that might benefit local inhabitants among others; but it is not possible at present to evaluate any potential impact or even speculate on the timetable for industrial projects.

Thus, we must turn to the farmers themselves and see what strategies they use to increase incomes and family security. Two main types of family strategies are possible and may be combined.

The first strategy is to send children to school in the hope that they will obtain nonagricultural jobs with good pay and contribute to the security of the family once they complete their schooling. According to our survey, 95% of the boys and 55% of the girls between the ages of 6 and 15 attend school, and 35% of the boys and 14% of the girls between 15 and 20 years of age. Parents attach great importance to boys' schooling, but have a tendency to plan matrimonial rather than professional futures for their daughters.

The second strategy is to invest in the farm in order to make it capable of producing more revenue. For this strategy to be successful, farms must be favorably located in regard to soil and water. In addition, farmers must have savings that they have accumulated through off-farm employment (often the case in Algeria, particularly in regions of high emigration), and/or access to loans through a network of social relations. There are also less tangible factors such as confidence in the future cohesion of the family (which is based on a prior investment in terms of traditional education), a positive attitude toward working the soil, and undoubtedly attributes of personality as well. This strategy may be manifested by the adoption of technical and/or economic innovations, buying or renting more land, or taking up poultry farming, growing potatoes, and cultivating irrigated fodder crops (an activity widespread in 1987-88 in reaction to the shortage of barley), as well as speculative behavior in response to opportunities offered by the market. A strategy of intensification can also result from entrepreneurs who calculate their investment with an eye on the expected profits. These individuals resort to the use of hired labor.

Recourse to hired labor is relatively widespread on the farms in our survey (Table 18). Even the small farms use hired workers, either for a few days of planting, or to pick vegetables to be marketed in a hurry. Most of the bigger farms employ hired labor on a permanent basis (tractor drivers, cowherds, etc.).

The jobs thus created are only of slight importance for the development of the region as a whole. However, they often provide opportunities for people on neighboring farms who need some extra income. These find themselves placed in a situation of classic dependent complementarity in relation to their larger and richer neighbors.

Table 18. Use of Hired Labor on Surveyed Farms by Farm Size Category.

Size Category (ha)	Number of Farms							
	Not hiring		0–15 days		10–50 days		50+ days	
	No.	%	No.	%	No.	%	No.	%
0-15	115	80	13	9	13	9	3	2
15-40	31	45	22	32	11	16	5	7
40+	8	22	6	16	4	11	19	51
Total	154	62	41	16	28	11	27	11

The complexity of concrete situations makes it hard to give an account of them by using quantitative indicators. Therefore, we will highlight the main features of certain noteworthy farms (the object of more detailed studies) that illustrate strategies in regard to the maximization of income and employment.

The small farms that we studied are relatively successful, given their small size, because of a combination of several integrated activities (production of fodder crops, use of fertilizer, labor by family members). They demonstrate the importance of the farmers' previous efforts augmented by off-farm earnings. All activities are oriented toward the market, with little or none of the final product consumed on the farms.

The medium-sized farms that we studied are characterized by the stability and cohesion of the families that reside on them. While they have consolidated their holdings and kept most family members on the farm, the owners of these farms have recently mobilized the family labor force in a quest for new cash revenues, adopting innovations and having recourse to bank loans as part of their strategy. They have not intensified cereals cultivation, but intensified farming by the introduction of new activities, such as vegetable growing or poultry breeding.

In the category of large farms, we find examples of farmers who practice the intensification of cereals cultivation (in one case, even without any associated livestock raising) in a carefully considered manner. All of them have a substantial array of agricultural equipment that they manage carefully. With one exception, they occupy large holdings of long standing, often reduced by the Agrarian Revolution and dating back to the colonial period. Their evident competence is shown in the rational conduct of operations, the reduction of costs, and knowledge of market conditions. They all have strong ties to town and city (through residence and/or work) and educate their children to secure good jobs.

To these individuals, who use old skills in the new technological and economic context, we can oppose another type of farmer who is not an heir, but who plays the game of the market, invests, gambles, and wins. He has abandoned the local cereals cultivation tradition in favor of raising livestock and garden produce on a relatively large scale. In contrast to the others, he puts into action a legitimacy granted by the present political system.

All these cases show that neither the traditional forms of social organization nor new political ideas contradict the development of a spirit of enterprise, but are in fact the necessary supports of it.

Conclusion

The owners of private farms in the two districts appear open to opportunities to change their practices in order to increase their revenues, without necessarily following the models proposed by specialists. This agricultural dynamism is expressed in the creation of new activities that lead to some new jobs for family members or hired workers. The manner in which the owners of large farms revise the local colonial model by combining cereals with sheep and adding market vegetables and poultry, while incorporating technological innovations that are within their reach (fertilizer, improved seed), is quite remarkable. The fact remains that these practices are reserved for the larger farms. The challenge for agronomists is not merely to increase the performance of the modernized farms but to set up systems of cultivation and models of production that correspond to the needs of the 'medium' and 'small' farmers under the difficult climatic conditions of a dry-farming environment.

References

Chaulet, C. 1987. La terre, les frères, et l'argent: Stratégies familiales et production agricole en Algérie depuis 1962. Algiers: O.P.U.

ONS. (Office National des Statistiques). Recensement general de la population et de l'habitat 1987, d'après la revue 'Statistiques,' 16 (juillet-septembre 1987). Algiers: Office National des Statistiques.

Perrin, R. Le Sersou, étude de geographie humaine. Méditerranée (avril-septembre 1960 & janvier-mars 1961): 33-93.

The Acceptance and Rejection of Agricultural Innovations by Small Farm Operators: A Case Study of a Tunisian Rural Community

ARBI BEN ACHOUR

Introduction

In most Third World countries, the national economy is dominated by the agricultural sector. However, agriculture is often unable to adequately feed the population and meet other fundamental national needs such as exports. Agriculture in such countries is often characterized by a dual structure: (1) a small modern sector using advanced technology and producing for the market, and (2) a large subsistence sector, often called 'traditional', using indigenous techniques and producing mostly for home consumption. For a variety of historical, economic, and political reasons, most of the rural population in the subsistence sector suffers from misery and poverty.

In order to increase the productivity of the traditional sector and improve social and economic conditions, governments and international agencies have tried strategies of development based on the adoption of new techniques of agricultural production. Many programs of agricultural and rural development have been implemented in the Third World. However, these programs have not solved the problem of hunger and poverty. On the contrary, there have often been unexpected negative effects such as (1) disruption of the farmers' socioeconomic organization (Griffin and Ghose 1984), (2) greater impoverishment among small farmers despite an increase in their productivity (Lofchie and Commins 1984), (3) increased inequality between large and small farmers (Schejtman 1984), and (4) most importantly, rejection of the proposed technologies. Indeed, studies have shown that small farmers reject new technology more often than not (Campagne 1978; Schejtman 1984).

This rejection has become a concern of international development agencies. Several research and development institutions are initiating research on the labor force/adapted technology continuum and on the satisfaction of the fundamental needs of small and medium-scale farmers. Which technologies have the best chances for adoption by these farmers? Which will most improve overall social welfare? What kind of information do they need to choose these technologies? What impact does transfer of technology have on the farm labor force?

These questions are especially relevant for Tunisia, where agriculture employs about half of the economically active population. More than four million

Dennis Tully (ed.), Labor, Employment and Agricultural Development in West Asia and North Africa, 165–189.
© 1990 ICARDA.

people live in rural areas. Therefore, it is crucial for Tunisia to develop a successful strategy of agricultural and rural development.

The most prominent objectives of the sixth Tunisian Economic and Social Development Plan are those objectives concerning an increase in agricultural production, attainment of national food self-sufficiency, creation of additional employment in rural areas, and an improvement in the standard of living. Tunisia has tried to meet these objectives by means of many rural and agricultural development projects. However, the results have been discouraging.

Indeed, these agricultural projects have not generally been able to meet the needs of the rural population, especially small farmers (GMLT 1973). Conscious of these problems, Tunisian public authorities have tried to reinforce development projects by the creation of special programs for small farmers. However, several evaluations of these programs (CNEA 1982) showed that they have not effectively improved the small farmers' conditions of life. Moreover, because of these programs, small farmers have become gradually more dependent on the market. Their traditional systems of production have been destroyed, with severe consequences such as erosion of the land, indebtedness, pauperization, and rural exodus (GMLT 1973).

Any relationship between the abundance of family labor and the adoption of new technologies can only be understood within the context of the farmers' environment. For example, one study (CNEA 1982) found that a combination of variables such as abundant family labor, very small farm size, production directed mainly toward family consumption, and low levels of farm income had a negative effect on the adoption of new technologies. However, on other farms, the combination of abundant family labor, production for the market, larger farm size (but still within the range of small farms), and favorable input and market prices had a positive effect on the adoption of new technologies. In both instances, there was an abundance of family labor, but farmers' use of new technologies was not the same.

The basic premise of this case study is that small and medium-scale farmers' attitudes toward innovations are strongly influenced by their circumstances. Farmers act and react according to their situation, objectives, goals and priorities, and orientation or attitudes. Thus, the adoption or rejection of new technologies is dictated by a number of interrelated facts or variables.

Situation refers both to the general situation of the total community (natural environment, history, social structure, culture) and to the farmer's personal situation as a member of that community (biography, family income, farm size, experience with extension agents, etc.). Farmers set priorities in order to make crucial economic decisions. Why does the farmer decide to produce a given crop? Is he producing for the market or for family consumption, or both? In each case, what kind of technology does he use and why? Finally, a logical reasoning proper to farmers dictates orientation or attitudes (e.g., acceptance or rejection of change). This line of logical reasoning is shaped by and takes into consideration (consciously or not) the farmers' priorities, personal situation, and general situation. In other words, both situations and priorities influence farmers' attitudes about the means to achieve their objectives.

Methodology

The purpose of this case study is to determine the factors associated with the use of labor and the adoption of new agricultural technologies in a small community, Lorbous, located in a semiarid area in northwest Tunisia (Map 1). This study includes an analysis of its history, community organization, and other relevant features.

In this community, there is an ongoing research project on systems of production, directed by several Tunisian agricultural research institutions. The combination of location and research constituted a unique opportunity to (1) tap an important but underutilized source of data, (2) synthesize the previous research, and (3) conduct a case study that would complement research activities in the area. Moreover, in Lorbous there was a farming system characterized by extensive cereal culture associated with livestock raising, as well as a predominance of small and medium-scale farmers, who had been exposed to previous technology projects introduced by government agencies.

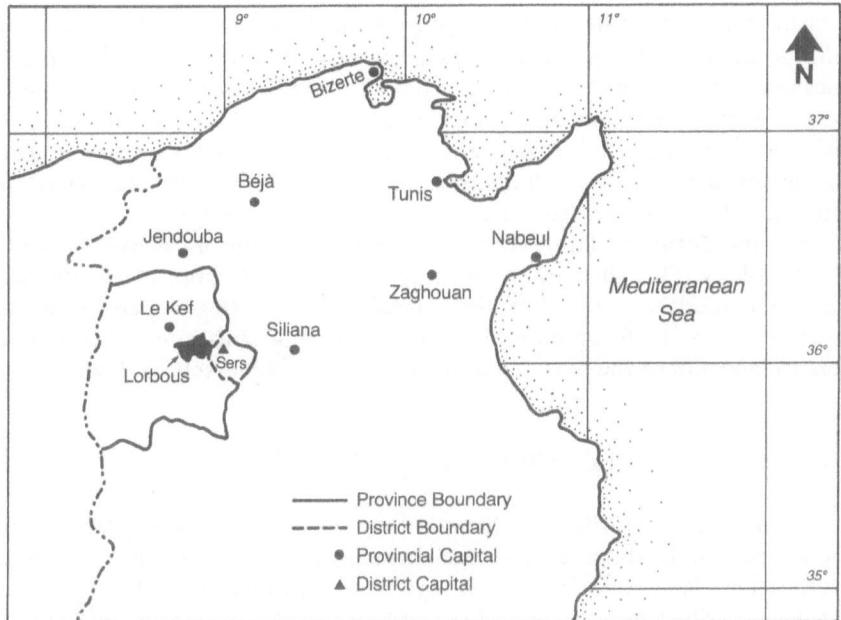

Map 1. Location of Lorbous.

Source: CNEA, Cartography Service (1987).

Table 1. Distribution of Farmers by Size of Farm in Lorbous and in the Sample.

Farm Size	Number of Farmers	
	Community	Sample
0 to 2 ha	99	33
2.5 to 5 ha	84	28
5.5 to 10 ha	58	19
10.5 to 20 ha	24	8
20.5 to 50 ha	12	5
Total	277	93

The village chief provided a list of farmers owning between 0.5 and 50 hectares in Lorbous. Size of farm was the main criterion used to select respondents for interviewing. This criterion is used by most projects involving small and medium-scale farmers in Tunisia. One-third of the population in each stratum was chosen randomly (Table 1).

I collected data and qualitative information from previous studies, research documents, and official documents. I also conducted in-depth interviews of key informants such as the chief of the village, influential elders, influential farmers, local officials, and researchers and scientists involved in past and present research and development projects in the area. Finally, I spent three months living with and observing farmers in their everyday life. I asked questions about matters such as socioeconomic status, family structure, involvement in previous technology transfer projects, and attitudes toward technological change. I interviewed a sample of twenty youths between fourteen and eighteen years old, asking them questions about how they view their future and how they perceive the modernization of agriculture.

From these primary data, I constructed scales and other measures of adoption of technology. This allowed me to compile variables and put them in multiple regression models to see how they related to the farmers' perceptions and attitudes vis-a-vis technological change and modernization. (For a more detailed account of the research and its results, see Ben-Achour 1988.)

Lorbous and Its Physical Milieu

The community of Lorbous is located in a basin-shaped valley (Map 2). The topography of Lorbous is hilly and broken, especially in the northeastern, southeastern, and western parts of the sector. In fact, only the central part can be described as a valley where cereals can be grown adequately. The valley is subject to continuous erosion due to the intensity of the rainfall, the type of soils, inappropriate land use, techniques of production, and the proximity of bare mountains or hills.

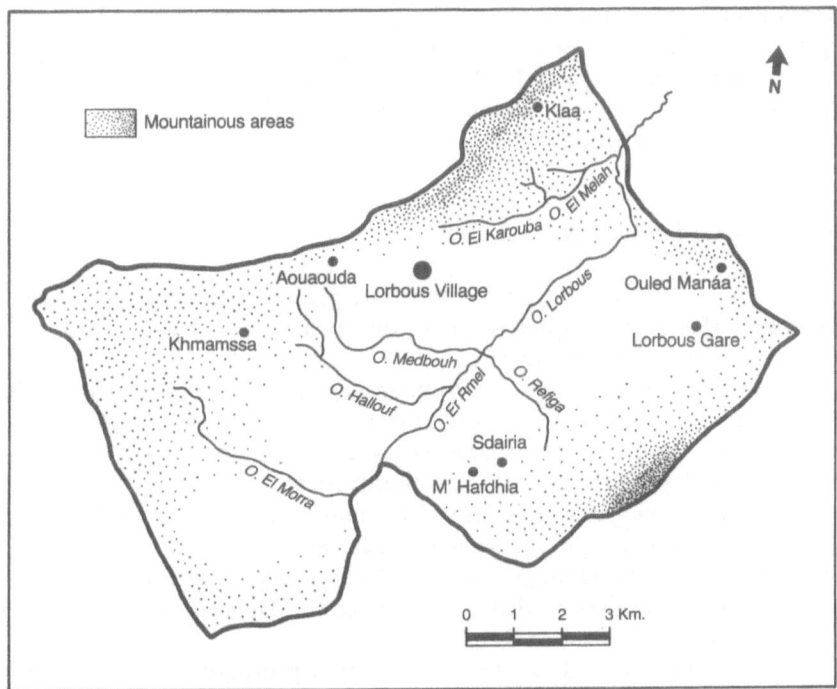

Map 2. Lorbous: Physical Setting.

Source: CNEA, Cartography Service (1987).

The annual rainfall normally varies from 400 to 500 millimeters a year. However, it is not rare to see as little as 250 millimeters, as was the case in 1985–86 (Table 2). The rain can fall very heavily in a very short period of time. The result is a powerful but irregular hydrographic network that contributes to the destruction of agricultural productivity by gradually washing away the arable soil. Year after year, the stream beds have become larger while new ones appear, causing further damage to the valley.

Table 2. Lorbous Monthly Rainfalls: 1985-86 Season.

1985	mm	1986	mm
June	0.5	January	25.0
July	3.0	February	55.0
August	0.0	March	77.0
September	4.0	April	19.0
October	18.5	May	16.0
November	15.5		
December	22.5	Total	256.0

Source: INAT (1986).

One of the biggest problems in Lorbous is the depletion of its soil resources. In general, the best soils (well-drained loam-silt soils) are located in the valley and are controlled by large farmers and a state cooperative. A large part of the land owned by small and medium-scale farmers is located in the sloping areas and is severely eroded.

The History Of Lorbous: A Brief View

The natural environment does not contribute positively to a highly productive agriculture. However, this hostile environment has not always existed. Much of the degradation is man-made.

According to a note prepared by Omda Abdelaziz Guezail (former village chief of Lorbous and father of the current chief), the city of Laribus was created by the Romans some 2,000 years ago, and once stood on the site of the present-day village and its surrounding areas.

Laribus was a key city that played an important military and economic role in the Roman Empire (Leo Africanus 1896; Cheikh Ibn Dhaif 1967). It was from Laribus and Dougga (another Roman city in northwestern Tunisia) that the Romans once controlled and defended the Tunisian cereal production that was gathered and sent to Rome.

Leo Africanus (1896) described Laribus as a prosperous city, famous for the fertility of its land, the abundance of its water, and the beauty of its surrounding woods. Under the Arabs, Laribus became Lorbous and played an important military role in pacifying the neighboring Berbers. In 800 A.D., the Aglabide dynasty chose Kairouan for its capital. Kairouan, however, was located in the arid center of the country and the surrounding plains were not good for cereal production. Consequently, the Aglabide rulers chose Lorbous as an important military, cultural, and economic center to control the northwest, which was so vital for their supply in cereals. Lorbous continued to prosper and played an important regional role until the Ottoman Empire took over Tunisia in 1574 A.D. The Ottomans decided to replace Lorbous with Kef as capital (because of its easily defended strategic position). Lorbous continued to be an important agricultural center. According to Omda Abdelaziz Guezail and other sources, before French colonization Lorbous attracted people looking for work during the harvesting period.

According to the local oral tradition, before the French colonial period food was never scarce for either humans or animals. Farmers were organized in extended families and land was used in common (undivided ownership).

When the French arrived in 1881, there were no property titles and land was considered public domain. Following this principle, the French authorities appropriated the best land and distributed it to French colonists. One thousand two hundred and sixty hectares of good land were taken away from the farmers.

The expropriated farmers were obliged to settle in the periphery of Lorbous (Map 2) and use marginal hillside land. The French introduced new techniques of production (mainly tractors, chemicals, and hired labor) along with monoculture (extensive cereal/fallow), all of which were soon adopted by large Tunisian farmers. Gradually, the small farmers were transformed into rural laborers. The colonists and the large Tunisian farmers needed a cheap and dependent labor force. Nothing was done to sustain the integration of those small farmers into the market economy. Such a situation, combined with the poor quality of the soil, the small size of the farms, and the demographic explosion, has contributed in large measure to the continuous pauperization of the small and medium-scale farmers.

When Tunisia became independent in 1956, agriculture was the largest sector in the national economy. But this agriculture was unable to feed the population and satisfy its fundamental needs. Except for some large farmers, the rest of the rural population was suffering from misery and poverty. The newly established government was compelled to adopt an agricultural and rural development strategy to help the small farmers. This strategy was built on the concept of modernizing the traditional structure of agriculture.

Given the inequality of land distribution and the predominance of the small farmers, those in charge of planning opted for the creation of cooperatives among these farmers. A government plan for social and economic development comments:

'An action of any depth must have an effect on the mental outlook of those concerned so that they can adapt themselves to a modern and healthy life with no place left for backwardness.... The social structure must be modified in order to permit the extremely poor social classes to have a better standard of living' (SEEN 1962, pp. 22-23).

In Lorbous, Mabrouka cooperative was created in 1962. Initially, the total size of the cooperative was 666 hectares. One hundred thirty-two small farmers joined or were forced to join the cooperative. The farmers contributed 365 hectares, with the remainder provided by the state. In 1966, 132 more farmers joined the cooperative, bringing in 896 additional hectares. The same year 205 hectares of forest were added to the cooperative, and a fusion with another cooperative brought an additional 275 hectares. In 1969, the cooperative incorporated more land and farmers, bringing its total size to 2,374 hectares, with 405 members.

The cooperative system did not improve the farmers' general conditions of life. Rather, they actually deteriorated. This strategy did not last very long; in 1969, President Bourguiba ordered a halt to the cooperatives. The failure of the cooperative system and its inability to reach its objectives are mainly explained by the authoritarian manner of the government in incorporating farmers into a system for which they had not been prepared, the lack of participation of the farmers in managing and operating the cooperatives, a rigid and dictatorial system of management through a centralized but ineffective bureaucracy, and bureaucratic contempt for the farmers' technical knowledge (Maklouf 1968; CERES 1968; Boulet 1971; Ben Achour 1976). Once the larger cooperative system was eliminated, the farmers were free to stay in cooperatives or to take possession of their land and farm it privately.

In Lorbous, the cooperative was dissolved in November 1969. Three hundred and ninety-three of the cooperative members left, taking with them 1,783 hectares to be operated privately. Two hundred and five hectares of hilly land also reverted to the public domain, leaving the cooperative with 891 hectares. Today the cooperative employs twelve workers. In the summer of 1987, I talked to six former members who told me, with bitterness, how miserable they had been, insisting that they had been very poorly remunerated, that their animals did not have the right to graze the land given to the cooperative, and thus had to be sold at cheap prices or slaughtered. One of them said, 'At the beginning we were willing to try it out, but after a while we discovered the real nature of the cooperative system, because we realized that instead of making a deal we were losing all we had. Sincerely, we were on the edge of starvation.'

A lot of money has subsequently been invested in Lorbous to help the farmers improve their systems of production. However, when it comes to development the real issue is not only what has been done, but how effectively, and how well it fits the needs of those for whom it was intended. The real questions that one ought to ask are: Did the Tunisian government learn anything from the cooperative experience? Were the farmers involved in the process of their own development, or were they consulted during the process? Did the officials, technicians, and researchers try to understand the farmers and their environment or did they consider them as simply objects that 'needed to be worked on?' How well organized and coordinated were the institutions involved and what means did they have to fulfill their duties? Were these development activities well supported by research?

It is by examining the issues raised above that one can begin to understand the farmers' behavior vis-a-vis the adoption or rejection of new technologies.

Lorbous, 2,000 Years after the Romans

In Table 3, I show the gross distribution of land use in Lorbous. The predominant crops are cereals, mainly durum wheat for the market and barley for animal consumption, followed by forage crops, primarily vetch and oats

Table 3. Land Use in Lorbous.

Land	Total Area	Barren Land	Agricultural Land	Arable Land	Forest and Pasture
Area (ha)	7,175	460	6,715	5,460	1,255
Percentage	100	6	94	76	17

Source: Direction Régionale de la Production Vegetale, Delegation du Sers.

Table 4. Use of Arable Land.

Land	Cereals	Legumes	Vegetables	Forage	Fruit Trees	Total
Area (ha)	4,763	30	30	341	296	5,460
Percentage	87.2	0.5	0.5	6.2	5.4	100

Source: Direction Régionale de la Production Vegetale, Delegation du Sers.

Table 5. Structure of Private Land Ownership in Lorbous.

Holding Size (ha)	Total ha	Percentage	Number of Owners	Percentage
0 to 2	340	7.5	99	34.4
2.5 to 5	310	6.8	84	29.2
5.5 to 10	431	9.5	58	20.1
10.5 to 20	363	8.0	24	8.3
20.5 to 50	451	9.9	12	4.2
Above 50	2,665	58.4	11	3.8
Total	4,560		288	100.0

(Table 4). However, some farmers have recently planted medic and sulla, which were introduced by a regional agricultural development agency (Office du Nord-Ouest).

At the national level, more than 40% of Tunisian farmers own 5.6% of the total agricultural land, whereas only 0.7% control 25.6% of the total agricultural land (Ommezzine 1982). In other words, 40.9% of the Tunisian farmers own, on average, 2.3 hectares, mostly located in dry farming areas.

In Lorbous, the situation is quite similar. The figures in Table 5 show clearly that there is land concentration in Lorbous. Moreover, among the eleven farmers that own 58% of the arable land, five are absentee owners. In contrast, the 99 farmers that survive in Lorbous by sharing 7% of the land individually own between 0.5 and 2 hectares. If one keeps in mind that agriculture in Lorbous depends heavily on the rainfall and is characterized by extensive cereal production and livestock activity which, to be successful, need a reasonable amount of land, it becomes extremely difficult to view these latter individuals as farmers. However, not being labeled farmers in the modern sense of the word (that is, making a living in producing for the market) does not mean that these people are not attempting to sustain a livelihood from the land. Farming for them is a way of life. It allows them to produce an important part of their own food, which reduces their dependence on the market, making it possible for them to accept occasional jobs, which in turn reduces the risk of rural exodus. Let us now examine the farmers' personal situations as revealed by the study.

The Farmers

In this study, I use the word 'farmer' to designate the person responsible for farm management decisions. All ninety-three informants are heads of households and thus responsible for the organization of the family labor force as well as the technical farm operations.

The farmer population is old, with a median age of fifty-five years and a modal age of sixty years (Table 6). Such an age structure might be predictive of a major disruption in the intergenerational transfer of farm operations within the rural population, and certainly indicates a rural exodus.

More than two-thirds of the farm operators interviewed cannot read or write. Of the third who have had some schooling, only one attended high school. Five attended Koran schools and the remaining two informants have a primary school education.

More than half of the farmers engage in some kind of off-farm work (Table 7). Nine have permanent jobs, six have their own small businesses, and the remaining twenty-eight have seasonal or occasional jobs. Ten of those who have an off-farm job are masons, nine are tractor drivers, and the rest are workers with no particular qualifications.

Of the thirty-nine informants who do not have off-farm work, one is retired. Only nine declared that they deliberately chose not to work because they are busy farming and taking care of their animals. Seventeen declared that

Table 6. Ages of Sample Farmers.

Age	Frequency	Percentage
30 years and younger	7	8
31 to 40 years	13	14
41 to 50 years	20	22
51 to 60 years	33	35
Above 60 years	20	22
Total	93	100

Table 7. Characteristics of Off-farm Employment.

Type of Job	Frequency	Percentage
Permanent job in agriculture in the area	15	28
Occasional job in agriculture in the area	12	22
Occasional nonagricultural job in the area	7	13
Permanent job in agriculture out of the area	4	7
Occasional job in agriculture out of the area	6	11
Small business in the area	10	19
Total	54	100

Table 8. Reason for Off-farm Employment.

Reason	Frequency	Percentage
Farming is not enough to make a living	39	72
To improve their general conditions of life	4	7
For more security for themselves and their families	3	6
Family needs are changing, they need more money to satisfy them	3	6
Money needed to invest in the farm to modernize it	1	2
Money needed to buy more land	1	2
Money needed to buy more animals	1	2
Farming is not enough to make a living, family needs are changing, and money needed to invest in the farm	1	2
Farming is not enough to make a living, family needs are changing, and money needed to modernize and buy more land	1	2

they were willing to work but could not find a job. Four are too old to work, and two have health problems.

Such a high percentage of farmers engaged in off-farm work (and/or seeking a job), and their reasons for doing so (Table 8), strongly support the argument that the agricultural sector's capacity to sustain its population is decreasing, making it less attractive to younger people.

Informal interviews (or discussions) with twenty young people between the ages of twelve and eighteen revealed that twelve of them are not willing to farm, nor are they willing to remain in Lorbous. In their view, farming is not feasible economically, and there is nothing attractive (except family ties for some) to keep them there. They mentioned the lack of good roads, youth clubs, running water, and electricity. Indeed, due to the new value system (modernism) introduced in the country since independence (1956), young people, even in the rural areas, have high aspirations and intend to reach objectives set according to these new values. Two-fifths of those who declared they are not willing to farm said that they could hardly wait until they were older to leave Lorbous and go abroad or to Tunis to find a good job. Of the twelve young people who are not willing to farm, seven dropped out of primary school, two are still in primary school, two are in secondary school and one dropped out of secondary school.

The remaining youths declared that they identify strongly with the land of their ancestors, but that it is hard to make a living because their holdings are too small. They indicated they definitely would farm if the government helped them buy more land.

Almost half of the farmers who are engaged in off-farm work are using the earnings from that employment to subsidize their farming operations, which indicates that these are economically not viable. Why are they staying on the farm? Perhaps they feel that the rest of the economy is so unstable that they could not find any better opportunities elsewhere. Or perhaps they fear

Table 9. Perception of Level of Income Compared to Average Fully-Employed Rural Worker.

Level of Income	Frequency		Percentage	
	Without*	With**	Without*	With**
By far inferior	9	7	10	8
Inferior	22	21	24	23
Equal	36	37	39	40
Superior	20	22	22	24
By far superior	6	6	6	6
Total	93	93	100	100

* Level of income without goods produced on the farm.
** Level of income with goods produced on the farm.

to leave a known situation with all its disadvantageous characteristics for an 'unknown' situation which, in addition to its economic uncertainties, may require formidable sociocultural adjustments.

In regard to level of income, I asked the farmers to suppose that a rural worker fully employed makes an average of 600 dinars (TND) per year (1 USD = 0.9 TND). Then I asked them to consider whether their annual monetary income was equal to, inferior to, by far inferior to, superior to, or by far superior to that of a full-time employed rural worker? After obtaining answers, I asked the farmers whether their level of income significantly changed when they added the goods produced on the farm and consumed by the family (Table 9).

In Tunisia, 600 TND a year is the income that will keep any citizen from falling into absolute poverty. (I obtained this figure by multiplying 2.4 TND, the minimum daily wage paid to rural workers, by 250 workdays per year. This amount is the same as the government figure of 50 TND per month.) It is significant to see that over 34% of the informants report their level of income falling below that line. When considering these figures, one must, however, be aware that people have a tendency to report less income than they really have; in cases where they truly don't know their exact income, they tend to perceive it lower than it really is.

Although self-subsistence still plays a crucial role in rural life, its importance seems to have diminished considerably. The needs of farm families have changed. Fewer goods are being produced for on-farm consumption, because of the migration of labor and changes in culinary habits, especially among the youth. A typical example is the preference by a large number of youth for 'commercial bread' to home-made bread, or fried tomatoes, pepper, and eggs to the traditional couscous. Such changes in preferential aspects of life styles make farmers more dependent on the market; yet they have little means available to allow them to be adequately integrated into the market economy.

Without off-farm jobs and livestock activities, the level of income would fall far below the line of absolute poverty, resulting in a greater rural exodus,

Table 10. Sources of Income in Order of Importance.

Classification	Frequency	Percentage
Off-farm activity, livestock, farming	29	31
Livestock, farming	27	29
Livestock, off-farm activity, farming	18	19
Off-farm activity, farming	11	12
Livestock, farming, off-farm activity	5	5
Farming, off-farm activity	2	2
Farming, livestock	1	1
Total	93	100

more mouths to feed, and more marginalization (Table 10). This argument is supported by the following results. For 53% of the informants, raising livestock, mainly sheep and goats, is the major source of income. For 44%, off-farm employment is the major source of income. Farming seems to play a minor economic role, due to the small size of the farms and the dispersal of the land.

The Farmers' Households

I use the word 'household' to designate all persons that live in the same house, and share meals and resources, including land.

The median household size is six: three adults and three children, with ranges of one to ten total family members, one to five adults, and zero to seven children.

In addition to wives and children, 17% of the informants have some relatives living with them (parents, brothers, sisters). This low percentage of other relatives living under the same roof clearly shows a gradual disappearance of the traditional extended family and its replacement by the nuclear family.

Thirty-eight percent of the children are under five years old. Thirty-eight percent are between 5 and 10 years old, and 23% are between 11 and 18 years old.

The percentage of children between 1 and 10 years old is high (77%). At that age, children can hardly be used as labor on the farm, although they can perform other domestic activities such as carrying water. Given the farmers' limited resources, the large number of children is an important financial 'responsibility.' This diminishes the farmers' capability to save money and improve the general conditions of life. Furthermore, poverty and numerous dependent children could have a negative effect on these children's education and their performance in school.

The level of illiteracy is even higher among the adult members of the households than it is for the farm operators. Sixty-nine percent are illiterate,

2% attended Koran schools, 28% have a primary school education, and 1% a secondary school education.

Seventy percent of the school-aged children are in primary school. Nine percent are in secondary school. Ten percent dropped out of primary school, and 11% (all female) were never sent to school.

Family Labor Force

Traditionally, rural families have chosen to have a greater number of children to secure their old age and to have enough labor to use on the farm. Today, in Lorbous, the latter expectations do not seem to be met.

To the question, 'Do the children help you on the farm?,' 45% of the informants who have children answered yes, and 55% answered no. Of the fifty-one farmers who answered no, four said that the children who are less willing to help are those who attend secondary school. Seventeen designated the children who dropped out of school and do not have a job. Nine designated the older children (between age fourteen and eighteen). Finally, twenty-one said that none of their children are willing to help on the farm. Farmers explained that the children are either not interested in agriculture, are being spoiled by new and modern values, or are simply lazy.

During the interviews with the twenty young people, five of them answered yes to the question, 'Do you help your parents do their agricultural work when they need it?' Seven said they help occasionally, and eight said never.

Four of those who said never explained their attitude by saying they are busy studying. Two said they prefer to look for paying jobs in or out of agriculture. The remaining two did not answer.

Such a drastic change in attitudes and the increasing drain of family labor toward off-farm jobs pose a problem. Indeed, it becomes irrelevant to try to estimate the importance of the availability of family labor on the basis of the size of the family. Rather, it becomes a matter of willingness and availability. Who knows better than the farmer who in his household is willing and/or available to help?

During the individual interviews, I asked the farmer to consider that a man can work on-farm 280 days a year and a woman 240 days a year when they are not employed elsewhere, 30 days when they are fully employed, and 60 days when they are employed part time. A child (between 10 and 18 years old) can work 30 days a year when enrolled in school, and 200 days a year when not enrolled and not employed. Then, for each member of his family, the farmer was asked to estimate the number of on-farm workdays per year each member provides or can (effectively) provide. The figures given for each member were added together then divided by an average of 240 workdays per year. The result was an index indicating the number of manpower units available in each household. The range varied from zero to four years (Table 11).

Table 11. Index of Manpower Units Available for On-farm Labor by Household.

Value	Frequency	Percentage
0	12	13
1	33	35
2	32	34
3	14	15
4	3	3
Total	93	100

One unit of manpower means that there are 240 days of work provided by members of the household to be effectively used on the farm. The number of units of manpower can be equal to or smaller than the number of members of the household, depending both on their willingness to help and their availability. For example, while one farmer, who has a family of ten members of working age can only count on one unit of manpower, another farmer with only three adult members in the family can count on three units of manpower.

The index shows that the number of manpower units available in each household for farm work is relatively low compared with family size. While 13% of the farmers cannot count on any family labor, only 3% can count on four manpower units. The channeling of labor toward off-farm jobs, along with a lack of interest in agriculture, are good indicators that farming activities will continue to decline unless existing conditions are changed.

Many of the farmers felt that the only young people who would stay in Lorbous were those who could not find jobs elsewhere, and 31% of the sample said that none of their children would stay. One gets the general impression that farming becomes the least attractive and last choice to be made when all other alternatives have been tried. Furthermore, the informants did not seem to be offended or shocked by the attitudes of the young people. Most believed that farming does not pay off and has no future, 73% said they would prefer their children to migrate, and 11% want to educate their children to get nonagricultural jobs. Farmers who preferred their children to migrate said that they could stay in Lorbous and farm only if more land, more jobs, water, electricity, and better services were available.

Whatever the questions asked of the farmers, their answers seemed always to be related to the lack of satisfaction in regard to some of their basic needs, namely, running water, electricity, public transportation, better roads, more jobs, and more land.

Crops and Forage Production

The main crops produced in Lorbous are wheat and barley. The most popular forages are vetch and oats. In Table 12 I classify the sample farms by type

Table 12. Classification of Farms by Crops.

Type of Production	Frequency	Percentage
Hard wheat, barley, vetch & oats	36	39
Barley, vetch & oats	13	14
Hard wheat, barley, vetch & oats, medic	13	14
Soft wheat, barley, vetch & oats	6	7
Soft and hard wheat, barley, vetch & oats	6	6
Soft and hard wheat, barley, vetch & oats, medic	5	5
Soft wheat, barley, vetch & oats	4	4
Hard wheat, barley, vetch & oats, chickpeas	3	3
Hard wheat, barley and medic	2	2
Hard and soft wheat, barley	2	2
Hard wheat, barley, vetch & oats, sulla	2	2
Hard wheat, barley, chickpeas	1	1
Barley	1	1
Total	93	100

of production based on the crop averages of the 1984–85, 1985–86, and 1986–87 seasons.

Durum (hard) wheat, barley, and a vetch-oats mixture are the most popular combination of crops and forage in Lorbous. Durum wheat is usually produced for the market, barley and soft wheat for both consumption (mainly livestock) and the market. About 18% of the farmers have tried medic. The adoption of sulla is slower, as only 2% of the farmers have tried it. Sulla (*Hedysarum coronarium*) is a perennial forage that, unlike medic, cannot be alternated with other crops. Most of the farmers have small holdings, and may refuse to plant a crop that doesn't become fully productive for two years.

Fallow plays an important part in the crop rotation pattern, which occurs biannually. Most of the farmers leave about half of their land fallow, mainly because of the importance of livestock in the area, but also to regenerate soil fertility. The farmers are not convinced that planting legume forages can replace this method of regenerating soil fertility without the loss of a food source for their herds. Such an attitude partly reflects a lack of confidence in the agricultural extension system.

New Techniques of Production

The use of tractors is becoming more common in Lorbous, even by the smallest farmers. However, mechanical plowing does not mean that the land is being prepared properly. Eighty-three percent of the informants who use tractors perform one deep and one superficial plowing. Seventeen percent perform one deep and two superficial plowings. Because they must hire tractors at

Table 13. Percentage of Farmers Using Techniques of Land Preparation.

Type of Operation	Percentage		
	84–85	85–86	86–87
Mechanical plowing	45	50	48
Traditional plowing	21	21	24
Combination of both	34	28	28
Total	100	100	100

a high hourly rate, farmers try to complete the plowing as quickly as possible. The dispersed fragmentation of the land is another handicap which keeps the land from being better prepared. Moving from one piece of land to another takes time that the tractor owners compute as labor time, which, given the farmers' low level of income, allows them little flexibility to spend more money to improve land preparation. Despite these problems, mechanization seems to be attractive to the farmers (Table 13). In cases where mechanical and traditional plowing are combined, tractors generally do the deep plowing, whereas farmers use draft animals for the second transverse plowing.

In regard to the use of high yield variety seed for wheat and barley (Table 14), the rate is high and relatively stable. The higher rate of use for wheat can be explained by the fact that wheat is mainly produced for the market. The 1985-86 low rate resulted from a severe drought.

Extension workers from both the Ministry of Agriculture and the Office du Nord-Ouest have promoted fertilizers. However, the use of chemical fertilizers does not seem to be very widespread (Table 15). Phosphorus and potassium are the most common. The rates of use are not uniform, but vary from one farmer to another according to their means, and from one season

Table 14. Rate of Use of High Yield Varieties (HYV).

Seed	Frequency			Percentage		
	84–85	85–86	86–87	84–85	85–86	86–87
Wheat	73	47	70	78	51	75
Barley	70	47	48	75	51	52

Table 15. Percentage of Farmers Using Chemical Fertilizers.

	84–85	85–86	86–87
No use of fertilizer	47	63	49
Use for wheat	21	11	19
Use for barley	7	2	3
Use for both	24	24	28

182

Table 16. Patterns of Herbicide Use by Farmers.

Patterns of Use	Frequency			Percentage		
	84–85	85–86	86–87	84–85	85–86	86–87
All crops	7	1	6	8	1	6
Small spot only	10	–	3	11	–	3
None	75	92	84	81	99	90
Total	93	93	93	100	100	100

to another according to the rainfall. The absence of adapted and standardized norms of chemical use in the area (and in the country) does not help to increase the usage.

The use of herbicides and pesticides is insignificant and is closely related to the belief that they burn the plants and make the animals (including the bees) sick when they graze the straws and stubble. Weeding is done by hand and the weeds are fed to the animals. During the three cultivating seasons, none of the informants used chemical pesticides.

The use of herbicides (Table 16) seems to be related to the availability of family labor because 86% (1984-85), 79% (1985-86), and 82% (1986-87) of the farmers who used chemical herbicides on their crops or on a spot to try them out had three or more manpower units. Spraying herbicide with manual pumps requires considerable labor.

Mechanical harvesting is becoming increasingly important (Table 17). During the harvest period, the family labor force, rather than working on the farm, seeks paid jobs available on large farms or state-owned farms. In 1985–86, only 21% of the farmers used mechanical harvesting techniques because of the severe drought.

Livestock

Raising livestock is a very important activity for the Lorbous farmers, as the sale of livestock contributes a substantial part of the on-farm income.

Table 17. Harvesting Techniques by Farmers.

Techniques	Frequency			Percentage		
	84–85	85–86	86–87	84–85	85–86	86–87
Mechanical	66	20	67	71	22	72
Manual	20	51	19	22	55	20
Both	7	4	7	8	4	8
Total	93	75	93	100	80	100

Fifty-six percent of the respondents own some cattle. Nevertheless, cattle are not an important component of livestock raising. Their needs for feed are greater than those of small ruminants. Even farmers who own more than twenty hectares do not have sizeable bovine herds. For most farmers, cattle raising is a secondary activity that provides milk and butter and a source of income when there is need for money.

On the other hand, all of the farmers interviewed, regardless of the size of their farms, own some sheep and/or goats. The sizes of the flocks range from one female to seventy head. The herds are much larger than the forage resources of the farms. Indeed, in the area on average, one hectare of vetch and oats forage produces 1,000 to 1,500 fodder units, one hectare of fallow produces 300 fodder units and one hectare of grazing land and/or stubble produces 250 fodder units (CNEA 1982). When compared to the 3,000 and 250 fodder units needed respectively by each cattle unit and sheep/goat unit, these resources could by no means satisfy the needs of animals unless the farmer bought additional feed, rented grazing land, resorted to over-grazing, or produced low-quality animals.

Farmers let their animals graze on almost every available spot, whether it be fallow land, cropland after harvest (straw), roadsides, or hills. This pattern of grazing is very damaging from an ecological and economic point of view. The pressure of animals on the land is too high (over-grazing). The straw could have been a good fertilizer. When the land is not plowed immediately after harvest, it hardens during the summer. If there is no rain in September, as is often the case, farmers find it very difficult to prepare the land properly. This has a negative effect on cereal yield.

Even though livestock activities are very important, the care of animals is poor. The interviews revealed that 92% of the informants do not use preventive medication for their sheep or goats and only 8% use it occasionally. For the cattle, the use of preventive medication is also very low. Livestock diseases can be treated free of charge. Incredibly, despite the fact that 81% of the farmers know this, very few of them use the veterinary services. The extremely low rate of use of preventive medication can partly be explained by the negative attitude of the farmers towards it, and by the failure of the agricultural extension services to provide adequate and reliable information.

Multivariate Analysis of Technology Adoption

The factors affecting the farmer's attitudes toward technological change are numerous and differ in importance from one farmer to another, according to specific circumstances. It would not be realistic to aim at resolving all of them at once. Thus, it becomes essential to identify the most important factors that affect attitudes toward technological change. We can do this by using multiple regression techniques.

We can state the hypothesis as follows: (1) farmers' attitudes toward adopting or rejecting new technologies are influenced by the combined (or simultaneous) effect of variables pertaining to their priorities and situation; (2) therefore, these attitudes cannot be explained by single variables; (3) consequently, the availability or unavailability of family labor *by itself* has very little (if any) effect on the acceptance or rejection of new technologies.

The assumption underlining this hypothesis is that farmers' attitudes toward new technologies are strongly influenced by their environment and their priorities. The dependent variable 'adoption of modern agricultural technologies' is an index constructed from the following practices: use of preventive veterinary medication, techniques of land preparation, use of high yield seeds, chemical fertilizers, chemical herbicides and pesticides, and harvesting techniques.

We assume that the number of items and types of innovation adopted indicate the farmers' attitude towards those items and operations. However, this raises a problem of applicability. For instance, if a farmer has no livestock, the use of preventive medication will not be applicable to his situation. Or if he does not grow wheat, the use of improved seeds for wheat will not be applicable. To avoid such biases, I constructed the index so that practices applicable to particular farms were weighted differently than those applicable to all farms.

The independent variables are either those repeatedly mentioned by farmers as being their major concern and/or those that refer to farmers' environment and priorities. These explanatory variables include number of units of manpower used and/or available on the farm, size of the farm, land dispersal, feeling about family well-being, confidence in public institutions, confidence in extension workers and technicians, income, age, education, and satisfaction with market prices.

Our objective is to make sure that, if there is a relationship between any one independent variable and the dependent variable, the variation in the dependent variable is not due to other interacting variables. Because determining the relation between availability of family labor and attitude toward technological change is an important goal, I checked to see if there is an interaction of labor availability with (1) farm size, (2) land dispersal, (3) level of income, and (4) level of satisfaction with the price system. Furthermore, because income and farm size are important factors to farmers, I also examined the effect of their interaction on behavior.

The results of the multiple regression analysis (Table 18) show a relationship between level of income and farm size and adoption of new technologies at the .05 level of significance (Alpha error = .0001). The R square indicates that 24% of the total variance in the adoption of new technologies is accounted for by the variance in income and farm size. The model shows also that income and farm size are interrelated. Therefore, we need to examine how their combined effect influences the farmers' attitudes toward the adoption of new technologies.

Table 18. Regression Analysis of the Effects of Farm Size and Income on Farmers' Attitude Toward Adoption of Agricultural Technology.

Variables	B	Partial	Significance
Interaction of income and farm size	−.013	−.272	.016
Level of income	.200	.459	.000
Farm size	.039	.276	.014

R square   = .24
Adjusted R square = .21
Constant   = .16

Table 19. Effect of Income and Farm Size on Adoption: Predictive Values of Adoption.

Farm Size	Income		
	Under 600 TND	600 TND	Above 600 TND
1 ha	.11	.30	.49
25 ha	1.90	1.76	1.63
50 ha	3.76	3.29	2.82

On small farms, the higher the level of income, the more likely is this level to influence the farmer's decision to adopt (Table 19). This relationship is indicated by the increasing predictive values for adoption, which slightly increase from .11 to .30 to .49 as the level of income increases when holding the farm size (1 ha) constant.

The figures in the second and third rows of Table 19 indicate that as the farm size increases and the level of income increases, the level of income is less likely to influence the farmer's decision to adopt new technologies. This relationship is indicated by the decreasing predictive values of adoption which decrease from 1.90 to 1.76 to 1.63 when holding the farm size constant at 25 hectares, and as indicated by the decreasing value of adoption from 3.76 to 3.29 to 2.82 when holding the farm size constant at 50 hectares.

Thus, if a farm is small, income is more important in affecting the adoption of new technology than if the farm is larger. Furthermore, there may be an economic threshold of either size or income or the combination of both that influences adoption.

This study found no evidence of a relationship between family labor force and the adoption of new technologies, even in interaction with other variables. In Lorbous, most farmers are integrated into the larger market economy. Their values have changed and so have their needs. The increasing need for money and market goods, along with the small size of their farms and their low level of income, forced the farmers and their families to sell their labor on the market rather than using it on the farm.

Rejection of Innovations

Most farmers want and look for changes that fit their needs. Technological packages are being offered to them as a means of change. In Lorbous, 72% of the respondents adopted certain technologies, then abandoned them, some after two cropping seasons (82%), and others for only one (18%) (Table 20). The greatest concern seemed to be with the high costs of chemical fertilizers, pesticides, and herbicides, relative to the low market prices for farm products, and the poor adaptability of chemical inputs.

Most farmers tried the innovations as a package and then rejected them. In a second step, the farmers adopted one or several items, but seldom the whole package. They modified the recommended quantities to better suit the environment according to their own judgment and experience. Only 14% of the informants adopted, then abandoned, new technology because it was in competition with their labor.

The farmers' behavior does not support the argument that they resist change because of their conservatism and backwardness. Conservatism certainly plays a role in the farmers' attitudes toward technological change, but this role interacts with other factors, making conservatism important in some cases and less important in others. Farmers responded to the question 'What is the basis for successful farming?' with a variety of answers including appropriate farm size, availability of advantageous loans, government support, availability of drinkable water, a good price system, and good services and infrastructure.

Only 11% of the farmers believed that successful farming consists of farming the way their ancestors did (which does not necessarily mean backwardness).

Over one-half of the farmers believed that development relies mainly on government assistance. Indeed, in-depth discussions revealed that most of them think of the state as a providential institution that should be doing everything for free. Such an attitude is the result of most of the assistance projects and programs put in place by the government, where farmers passively receive assistance (not development) with very little participation. Such projects and

Table 20. Innovations Tried Then Rejected by Farmers.

Technologies	Frequency	Percentage
Fertilizers, pesticides & herbicides	43	57
Pesticides & herbicides	27	36
Fertilizers	2	3
Pesticides	2	3
Herbicides	2	3
High yield seeds, mechanical plowing & harvesting	2	3
High yield seeds, mechanical plowing & fertilizer	2	3
Total	76	100

programs, instead of helping the farmers, create in them an outlook of dependence and diminish their sense of responsibility.

The small farmers often find themselves poorer as a result of the projects. Consequently, they don't often follow the advice of change agents. They have learned from experience that the technologies proposed may increase their production. But they also increase expenses. The market prices and increased production often don't compensate for higher costs.

Conclusion

There are inherent contradictions between national food policies and initiatives to improve the standard of living of small farmers by increasing their income and creating employment. The government intends agricultural programs to be a means to increase the production of nutritious, cheap food for the urban areas, or improve the balance of payments by increasing exports and decreasing imports. For these reasons, the public authorities offer low prices for the farmers' products. On the other hand, farmers want to satisfy their fundamental needs for adequate food, decent housing, drinking water, electricity, health, and education. To achieve these, they try to avoid the use of expensive technologies and inputs to reduce risks and, most importantly, aspire to capture a larger part of their surplus value by means of a good pricing system for their products and a low price for inputs. Therefore, any project that is not directed toward the satisfaction of these needs will likely fail.

The conclusion drawn from these contradictions is that the national objective of increasing agricultural productivity from small farms will not be attained unless the fundamental needs of the farmers are satisfied.

But with such small farm sizes, it is difficult to successfully attain the desired objectives of development. In Lorbous, the use of new technologies has increased farm productivity. But the small size of the holdings and the irregularity of rainfall render the principle of 'economy of scale' not applicable, making the use of these technologies not profitable for the farmers.

Indeed, the statistical analysis showed that level of income and farm size were the most important factors that affect the farmers' adoption of new technologies. It further showed that when the farm is small and the level of income is relatively high, it is more likely that farmers will adopt new technologies. When farm size is greater and the level of income is higher, it is farm size that primarily influences a farmer's decision to adopt. This means that because farm size is greater, its profitability increases, and the farmer will try to use techniques of production that help to sustain this profitability. When the farm size is very small, its profitability is insignificant, resulting in a low level of income and a low level of adoption of technological innovations. The small farm size also reduces the need for family labor.

The farmers' personal and general situations also affected their attitudes toward adopting or rejecting new techniques of production. Indeed, the socioeconomic environment is not favorable to the farmers: land is too dispersed, and there is poor housing, lack of water, inadequate public services, and poor infrastructure. The low level of confidence of the farmers in public institutions and services, and the equally low level of satisfaction with their own lives, are expressions of this negative situation.

The overall conclusion from this study is that attitudes toward the adoption of new agricultural technologies are not directly affected by the availability of family labor, but are related to a multitude of factors, the most significant being the size of the farm and income. If genuine development is to be sustained in Lorbous, farm size will have to increase, at least for the smallest farms.

Acknowledgements

My special thanks go to Dr. Rex Campbell, Chairman of the Department of Rural Sociology at the University of Missouri-Columbia, for his continual support, assistance, and advice. My thanks also go to Professors Daryl Hobbs and Joel Hartman for their valuable assistance. I am also grateful to INAT (Institut National Agronomique de Tunisie), especially Dr. Ali Salmi for the unconditional support I received during fieldwork. Finally, the continual support provided by Omda Mokthar Guezail during fieldwork is deeply appreciated.

References

Ben-Achour, A. 1976. Le reforme agraire en Tunisie: 1962–1969, L'anti democratie dans les UCP et leur mechanisation à outrance, facteurs principaux de son echec. EHESS, Sorbonne, Paris.

Ben-Achour, A. 1988. An Enquiry into the Basis of Rejection of Agricultural Innovations by Small Farm Operators: A Case Study of a Tunisian Rural Community. Ph.D. dissertation, University of Missouri, Columbia.

Boulet, D. Etude economique des cooperatives agricoles de production en Tunisie. Options Méditerranéenes (April, 1971): 97–103.

Campagne, P. 1978. Analyses des zones rurales: Elements pour une approche paysanne. IAM, Montpellier, France.

CERES (Centre d'étude et de Recherche Economique et Social). 1968. Revue Tunisienne de Science Sociales, 12 (January 1968). University of Tunis, Tunis.

Cheikh Ibn Dhaif. 1967. (History of Tunisia and the Aglabide Dynasty.) In Arabic. Tunis: La Maison Tunisienne de l'Edition et de la Diffusion.

CNEA (Centre National des Etudes Agricoles). 1982. Projet de développement des petites et moyennes exploitations dans le Gouvernorat du Kef et Siliana. Tunis: CNEA.

GMLT (Groupe Marxiste Leniniste Tunisien). 1973. Contribution a l'étude de la question agraire en Tunisie. Mimeo.

Griffin, K., & Ghose, A. 1984. Growth and impoverishment in the rural areas of Africa. Pages 242–273 in The Political Economy of Development and Underdevelopment (3rd edition, C. Wilber, Ed.). New York: Random House.

INAT (Institut National Agronomique de Tunisie). 1986. Rapport annuel des activités de recherche 1985-86. Projet de recherche agricole sur les systemes de production. Tunis: INAT.

Leo Africanus. 1896. The History and Description of Africa and of the Notable Things Therein Contained, Done into English in the Year 1600 by John Pory. (3 volumes, R. Brown, Ed.). London: Hakluyt Society.

Lofchie, M., & Commins, S. 1984. Food deficits and agricultural policies in tropical Africa. Pages 222–241 *in* The Political Economy of Development and Underdevelopment (3rd edition, C. Wilber, Ed.). New York: Random House.

Maklouf, E. 1968. Structures agraires et modernization de l'agriculture dans les plaines du Kef. Tunis: Cahiers du CERES.

Ommezine, A. 1982. An Economic Appraisal of Agricultural Cooperatives in Tunisia. Ph.D. dissertation, University of Missouri, Columbia.

Schejtman, A. 1984. The peasant economy: Internal logic, articulation, and persistence. Pages 274–298 *in* The Political Economy of Development and Underdevelopment (3rd edition, C. Wilber, Ed.). New York: Random House.

SEEN (Secretariat d'Etat a l'Economie Nationale). 1962. Le plan triennal (1962–1964). Tunis: SEEN.

Farm Mechanization and Socioeconomic Changes in Agriculture in a Semiarid Region of Tunisia

ALIA GANA and RAOUDHA KHALDI

Introduction

The general use of mechanized farm equipment for plowing and harvesting operations is one of the most remarkable developments that Tunisian agriculture, particularly rainfed agriculture, has experienced since independence. While in 1961-62 only 18.2% of the farmers used tractors (SEA n.d.), by 1980 the rate had increased to 56.6% (MA 1981). On small farms under twenty hectares, the rate of tractor use increased from 13.2% in 1961–62 to 52.2% in 1980 and on medium-sized farms from 38.2% to 77.7% (MA 1981).

Mechanization of farm operations has been one of the main factors for the increased production of large-scale rainfed farm crops. Since independence, mechanized farm equipment has received a high rate of investment. From 1973 to 1981, farm mechanization received 25% of all investment in agriculture (MA 1981).

During the 1960s, the movement toward mechanization gained ground principally among agricultural cooperatives formed by the government. Development planners believed that only the existence of large-scale farms would allow for a rapid increase in mechanization and, consequently, a substantial increase in production. After the failure of the cooperatives in 1969, investment efforts focused on farmers in the private sector, who were encouraged by a low-interest credit policy to mechanize their operations. Even though the credit facilities benefited well-to-do farmers, the long-term effect of this investment policy would be felt throughout the agricultural sector as all categories of farmers gradually acquired access to rented tractors and combines.

Moreover, this progressive mechanization movement would have major effects on farm labor. The rate of employment in agriculture dropped from 68% of the total active population in 1956 to 31% in 1980. This decrease was of course also due to rural-urban migration and the development of employment opportunities in other sectors of the economy. In absolute figures, the total active farm population decreased from 859,304 in 1956 to 562,100 in 1980 (INS 1984). In addition, figures published by the Ministry of Agriculture indicate a major decrease in the number of full-time wage-earning farm workers (a drop of 27.4% between 1969 and 1980) (MA 1981).

The increased use of household labor represents one of the essential features

Dennis Tully (ed.), Labor, Employment and Agricultural Development in West Asia and North Africa, 191–214.
© 1990 ICARDA.

in the development of Tunisian agriculture. Another characteristic is the great number of part-time farmers (43% of all farmers according to the agricultural survey of 1985) (MA 1985). Part-time farmers are mainly those who are needy. The rate of part-time farmers on farms of less than 5 hectares is 57% while it is only 18% for farmers who work farms of more than 100 hectares. We can thus explain part-time farming by the necessity for certain farmers to supplement the small income their farms provide with other sources of income.

The increasing number of small part-time farmers seems to correspond to a marked increase in the number of women now working on farms. According to census data, the active female population increased by almost 64% between 1975 and 1980, while the male population remained stationary during the same period (INS 1984). The increasing number of female workers who are today becoming a feature of the agricultural sector (as seasonal, low-paid laborers, or replacements for farmers who have other jobs) seems to reflect a certain alienation and devaluation of farm labor.

If all of these changes in the basic characteristics of farm labor have not been exclusively caused by farm mechanization, we cannot deny that a certain number of them have, and that the movement which has made the tractor a means of production as important as the land itself has completely upset traditional agriculture in Tunisia.

The purpose of this case study is to examine the process of change resulting from mechanization in a small rainfed cereal-growing zone. First, we will analyze the factors that determine the adoption of farm mechanization in terms of the conditions that prevail on different types of farms. We hypothesize that a farmer's selection and use of technology depends on a number of factors related to the farm itself, the socioeconomic environment, and the sociocultural models or value systems of the farmer.

Second, by identifying differences among types of farms, we will examine the socioeconomic consequences of mechanization on farm management, labor, production, and socioeconomic relationships.

Finally, we will formulate proposals for policies likely to improve technological development in agriculture. One of our objectives is to determine the requirements of a policy for technological innovation that is better adapted to the social and economic realities of the farms, as well as to those of national development as a whole.

Methodology

The zone chosen for the purpose of this study includes three subdistricts, two in the district of Goubellat and one in an adjacent district of Zagouan province (Map 1). The subdistricts are located in a semiarid region on a wide plain surrounded by mountains. This plain is divided into two topographical zones: a nonarable zone used for forest and pastures, and an arable zone for cereals cultivation. Rainfall is irregular, with an average of some four

Map 1. The Study Area.

hundred millimeters annually. The dry season can last from five to eight months each year.

Small farms of less than 10 hectares account for 70% of all farms while those of more than 50 hectares represent 3% of the total. The large farms occupy over 50% of the total surface area, with the rest divided between small and medium-sized farms. All farms, regardless of size, are oriented toward single-crop cereals farming. Cereals cover more than 50% of the total utilized agricultural area, with durum wheat predominant. On small farms, the farmers cultivate cereals along with certain vegetable crops used for human consumption or animal feed. Forage crops are cultivated on 6% and food legumes on 2% of the total utilized agricultural area. Because these crops are not yet organized into a precise rotation system, the association between livestock, forage, and cereals is still inadequate. There is extensive livestock raising, with the use of grazing lands and areas of stubble. Under current soil and climatic conditions, grazing lands account for some 30% of the total utilized agricultural area. This rate is somewhat lower on the small farms, which diversify their crop systems more than large ones. The use of fertilizers depends on climatic conditions and is inadequate and insufficient.

Even though household labor is available in the region, there is a certain mobility of labor on and off the farm. Some 45% of the farmers have other jobs, usually as full-time farm workers. Farms with more than thirty hectares have tractors that can modify cultivation methods and increase the degree of crop intensification.

We surveyed a sample of twenty-six farms typical of the diversity of farming structures and systems in the region. We selected these farms from the 572 farms in the region that had been studied in a previous research project, whose results were available to us (Projet de coopération INRAT/INAT/ICARDA/ CRDI sur le système de production dans la zone de Goubellat). We based our sample choice on two major criteria: farm size and mechanization (possession of tractors). Our main purpose in the survey was to identify the determining factors (social, economic, technological) and the effects of mechanization on farm management and development. We also undertook complementary studies at the local and national levels. At the local level, we carried out inquiries with heads of farm services for a better understanding of the problems of farm mechanization (availability of machinery, problems of supply, relations between owners and renters, etc.), as well as relations between farms and their social, technological, economic, and political environments. At the national level, we compiled a bibliography of case studies to identify problems related to farm mechanization throughout the country.

The resulting data were subjected to computer analyses to discover correlations between variables, both for individual farms and farm groups.

General Characteristics of Sample Farms

We classified the farms in this investigation into four groups according to two major criteria: farm size and ownership of farm machinery (Table 1).

The average age of farmers in the sample is 59, which is the average age of farmers at the national level. A large majority (65%) are over 50 years of age. Only 8% of all farmers in the sample are under 40 years of age. Thirty-five percent of the sample are illiterate; 47% received some instruction in Koran schools, and 24% attended public schools (12% primary and 12% secondary). For the most part, the latter belong to Group 4 (the largest farms).

The vast majority of farmers (85%) come from families that own farmland while 4% come from families of farm workers. Most of the farmers (81%)

Table 1. Farm Distribution by Group.

Group	Farm Size (ha)	Farm Machinery	Number of Farms	%
1	<10	none	7	27.0
2	10–30	none	8	30.5
3	30–100	tractor	4	15.5
4	>100	tractor and other	7	27.0
Total			26	100.0

live on their own farms, and most nonresident farmers are large property owners.

Some 67% of the farmers have at one time or another worked away from their farms. At present, 27% of the total hold off-farm jobs, either as laborers (43%), or in commerce or the civil service (43%). The highest percentage of part-time farmers is found in Group 2 (farms under 30 hectares).

Farm families have an average of six children, and 54% of the families in the sample have seven or more. The average number of residents per farm is eight and the average number of dependent persons is seven. However, 57% of the farms have more than eight residents, and 54% of them more than seven dependents.

Most of the farmers (77%) have no water sources on their farms; on the average, family members must go at least four kilometers to obtain water.

Our findings indicate that 69% of the farmers work their own land. Fifteen percent of these farmers also work other lands in association with family members. Working land in association with others seems to be a strategy for the extension of cultivated area. The land in use by farmers in our sample varies from 3 to 400 hectares, with the average farm being 5.5 hectares in Group 1, 18.3 hectares in Group 2, 45.3 hectares in Group 3, and 199 hectares in Group 4. A high percentage of the farms in the region under study are worked by the landowners themselves. Thus, 46% of the farmers in our sample

Table 2. Distribution of Land Tenure Among Farmers in Each Farm Group.

	Group 1	Group 2	Group 3	Group 4	Total (All Groups)
Owns land					
Number of farmers	7	7	4	7	25
Percentage of group	100	88	100	100	96
Area owned (ha)	30	115	158	849	1150
Percentage	79	77	78	61	64.5
Rents land					
Number of farmers	1	2	2	3	8
Percentage of group	14	25	50	43	31
Area rented (ha)	3	26	23	282	334
Percentage	7.8	18	11	20.3	19
Sharecrops land					
Number of farmers	1	2	2	5	10
Percentage of group	14	25	50	71	38
Area sharecropped (ha)	5	8	22	263	298
Percentage	13	5	11	19	17
Total (Farmers)	7	8	4	7	26
Total (ha)	38	149	203	1394	1783

work only their own land and 96% of them own all or part of their total cultivated area. The percentage of farmers who farm only their own land is inversely proportional to the size of the farms. It drops from 71% in Group 1 to 14% in Group 4. The percentage of farmers who are sharecroppers is slightly higher than the percentage of those who farm rented land (38.5% as compared to 30.5%). The majority of sharecropping contracts (57%) are third party contracts. The average rental price per hectare is 33 dinars (TND) with a registered maximum of 50 TND (1 USD = 0.9 TND). In Groups 3 and 4, there is a higher proportion of farmers holding land either for sharecropping or for rent.

In terms of cultivated area, the privately-owned lands represent 64.5% of the total utilized land surface in the zone under study (Table 2). In all groups, the area cultivated by owner constitutes the major portion of the farm holdings. This area tends to decrease in the large farms of more than 100 hectares (61% of the cultivated area as opposed to 79% for farms in Group 1). The highest proportion of land area rented or worked in sharecropping occurs on farms in Group 4.

Crop Production

Cereals are the major crop for all farms in our sample. Half of the farmers cultivate cereals in association with fodder crops, food legumes, and tree crops. All of the farmers combine crop production and raising livestock (Table 3).

In 1986–87, farmers cultivated cereals on 48% of the total utilized agricultural area (barley on 58% of this area, durum wheat on 33%, and soft wheat on 9.5%). Most farmers (89%) grow barley and wheat (Table 4). In Groups 1 and 2, areas cultivated in durum wheat and barley are more or less equal. Soft wheat accounts for only 2.5% of the total cereal area on these farms. Some 31% of the farmers cultivate soft wheat; the number of hectares planted in soft wheat increases according to farm size. Barley is predominant on Group 4 farms where it represents some 62% of the total area cultivated in cereals. The average yield of durum wheat is 1260 kg/ha and varies according to group from 980 kg/ha (Group 1) to 1790 kg/ha (Group 4). The average yield for soft wheat is 1200 kg/ha and for barley is 1450 kg/ha (2070 kg/ha in Group 4).

All farm families reserve a certain quantity of durum wheat for their own consumption. In Group 1, 53% of the production is kept for the household. The proportion of durum wheat retained for household consumption is inversely proportional to farm size. In farms 100 hectares or larger, 15% is reserved for this purpose. Farmers grow barley for sale or for fodder. On all the farms, more than 50% of the production is sold (85% in Group 4). Farmers in Groups 1, 2 and 3 use large amounts of barley for fodder.

Forage crops follow cereal crops in order of importance. They occupy some 17% of all cultivated areas and are grown by 73% of the farmers in the sample.

Table 3. Crop Production Systems by Farm Group.

Groups	With Tree Crops				Without Tree Crops			
	Cereals Only	Cereals & Forage	Cereals & Legumes	All Three	Cereals Only	Cereals & Forage	Cereals & Legumes	All Three
Group 1								
No. of farmers	0	2	1	3	0	0	0	1
Percent of group	0	28.5	14.2	42.8	0	0	0	14.2
Group 2								
No. of farmers	1	0	1	4	0	1	0	1
Percent of group	12.5	0	12.5	50.0	0	12.5	0	12.5
Group 3								
No. of farmers	0	0	0	3	0	0	1	0
Percent of group	0	0	0	75.0	0	0	25.0	0
Group 4								
No. of farmers	0	3	0	3	0	1	0	0
Percent of group	0	42.8	0	42.8	0	14.2	0	0
Total (farmers)	1	5	2	13	0	2	1	2
Percent of total	3.8	19.2	7.6	50.0	0	7.6	3.8	7.6

Table 4. Total Crop Area by Farm Group and by Crop.

Crop		Group 1		Group 2		Group 3		Group 4		All Groups	
		Farms	Area (ha)	Farms	Area (ha)	Farms	Area (ha)	Farms	Area (ha)	Farms	Area (ha)
Durum wheat:	No.	6	9.5	7	41.0	4	35.0	6	160.0	23	245.5
	%	86	21	88	27	100	19	86	13	89	16
Soft wheat:	No.	1	0.5	2	2.0	2	13.0	3	55.0	8	70.5
	%	14	1	25	13	50	7	43	5	31	4
Barley:	No.	5	10.0	7	37.0	4	32.0	7	354.0	23	433
	%	71	22	88	24	100	18	100	30	89	28
All cereals:	No.	6	20.0	7	80.0	4	80.0	7	569.0	24	749
	%	86	44	88	53	100	44	100	48	92	48
Forage:	No.	4	7.3	5	15.0	3	16.0	7	227.0	19	265.3
	%	57	16	63	10	75	9	100	19	73	17
Legumes:	No.	3	0.5	5	4.9	4	13.0	3	33.0	15	51.4
	%	43	1	63	3	100	7	43	3	58	3
Tree crops:	No.	6	11.8	6	14.8	3	8.8	6	71.0	21	106.4
	%	86	26	75	10	75	5	86	6	80	7
Fallow:	No.	2	5.5	7	36.9	3	63.5	7	294.0	19	400
	%	29	12	88	24	75	35	100	25	73	24
Total	No.	7	45.1	8	151.6	4	181.3	7	1194.0	26	1572.1

The number of farmers who cultivate forage crops increases in proportion to the size of the farm (57% in Group 1 and 100% in Group 4). As for land area, the largest quantities of fodder crops are grown on large farms where 19% of all cultivated land is used for this purpose. These crops are principally vetch and oats. Vetch is used for animal feed in Groups 1, 2, and 3, while in Group 4, 87% of the crop is sold on the market.

Food legumes are cultivated by 58% of the farmers and cover 3.2% of the cultivated area. All farmers in Group 3 cultivate legumes, while in Groups 1, 2 and 4 the rate is somewhat lower. In regard to land area, legumes are more widely cultivated on Group 3 farms (7.2%). These farms have available household labor and a tendency to diversify their production. Faba beans are the most commonly cultivated legumes; 46% of the farmers in the sample (especially in Groups 2 and 3) grow them, mostly for sale. They are an excellent cash crop for those farmers who work small and medium-sized farms. They can also be used for human and animal consumption on the farm.

Eighty percent of the farmers cultivate tree crops on 7% of the land area. The main species is the olive tree, which grows on all types of farms. Production is for household consumption as well as for the market. On the small farms, farmers cultivate cereals, forage crops, and food legumes on the same plots of land where olive trees grow. This intercropping enables farmers to compensate for a limited land surface.

In the research zone, 25% of the total utilized agricultural area consists of fallow land. The number of hectares left in fallow is the lowest on Group 1 farms (12% of the cultivated area). In other groups, it varies between one-fourth and one-third of the area under cultivation. According to farmers, the land must be allowed to rest and provide grazing for animals.

Crop Techniques

Farmers use cereal seed from their own harvest or else purchase it, either from other farmers or the Cereals Board (Office des Céréales). They can also exchange their seed for improved seed from seed cooperatives or the Cereals Board. Yet only 12% of the farmers exchange their durum wheat seed every year, and 48% every two years. Twelve percent of all farmers never exchange their durum wheat seed (among farmers in Group 1, the rate reaches 33%). For barley, the percentage of farmers who use only their own seed is particularly high (84%), because of the scarcity of improved varieties for barley.

In contrast to improved seed, the use of fertilizers is widespread, especially phosphorus fertilizer for cereals. However, 43% of the farmers in Group 1 cannot afford to use either phosphorus or nitrate fertilizer, which is applied less systematically. For cereals, the percentage of farmers who use nitrate fertilizer is 54%.

The use of chemical herbicides is even less than that of phosphorus and nitrate fertilizers. Only 77% of the farmers use chemical herbicides on a systematic basis. This rate reaches 43% for farmers in Group 4. On the other hand, 71% of the farmers in Group 1 and 63% of the farmers in Group 2 never use herbicides. Farmers state the following reasons, in order of importance: lack of money to purchase herbicide, grazing requirements, absence of weeds, and the toxic effects of herbicide on olive trees when cereals are intercropped.

Most of the farmers in our sample (85%) practice a three-year crop rotation of fallow-wheat-barley or forage. A few farmers in Group 2 (4%) practice a four-year rotation (legumes-wheat-barley-forage). Eleven percent of the farmers in Groups 1, 2, and 4 observe a biennial pattern (fallow-wheat). In Groups 2 and 4, this two-year rotation alternates with a three-year rotation.

Livestock

In the research zone. crop production is closely linked with raising livestock. Almost all of the farms (97%) have sheep, except for a few in Group 1 (the smallest farms). The total number of sheep in the flock varies greatly from one group to another (Table 5). In Group 1, 71% of the farmers have fewer than 10 sheep, while the percentage of farmers with fewer than 10 sheep drops to 13% in Group 2 and zero in Groups 3 and 4. In Groups 2 and 3, the total number of sheep per flock ranges from 10 to 50, while in Group 4 it usually exceeds 100.

Most of the farmers (80%) own cattle. In Group 1, all farmers raise cattle but rarely more than three. In Groups 2 and 3, the total number owned per farm ranges from two to five. Only in Group 4 can one find herds with as many as 50 animals. The vast majority of purebred dairy cows are also in Group 4 (94% of the herd).

Livestock receive nourishment from farm produce, purchased animal feed, or grazing on rented pasture. The amount of feed produced on the farm increases with farm size. On the other hand, grazing cattle on rented pasture occurs more frequently on small farms.

On all categories of farms, farmers produce meat (lamb and veal) to be

Table 5. Percentage of Farms Raising Sheep.

Number of Sheep	Group 1	Group 2	Group 3	Group 4	Total % (All Groups)
0	28.5	0	0	0	7.6
1–4	28.5	12.5	0	0	11.5
5–9	14.2	0	0	0	3.8
10–49	28.5	62.5	50.0	14.2	38.5
50–99	0	12.2	25.0	14.2	11.5
100+	0	12.5	25.0	71.4	26.9

sold on the market. Milk, however, is consumed by the household except on those farms with large herds (Group 4).

Equipment

Farmers in our sample rarely use draft animals. They are found mostly on small farms where they are used for transportation and, to a lesser extent, for farm work on inaccessible plots of land or on uneven, rough terrain.

Farmers in Group 4 own 12 of the 16 tractors in the sample, 3 of the 4 seeders, and 6 of the 7 fertilizer spreaders, as well as the only 3 combines, 2 balers, and 1 mower. The other equipment is owned by Group 3 farmers. In Group 4, there can be as many as 3 tractors per farm, as on 29% of the farms in this group. On Group 3 farms, the farmers have on the average, one tractor for 50 hectares, while on Group 4 farms there is one tractor for every 116 hectares. The farmers in Group 3 often rent out their equipment to other farmers at a much higher rate than do those in Group 4 (75% as compared to 50%).

Labor

In Groups 1 and 2, farm work is performed primarily by household labor, namely the farmers, their wives, and their children. The use of household labor is also important in Group 3. Thirty-one percent of the sample farmers never employ paid workers. These belong to Groups 1 and 2. Resorting to paid part-time labor is fairly frequent in Group 3. Most of the farms (69%) use part-time paid labor while full-time paid workers are employed on only 23% of the farms. Most of these farm workers (95%) are found on Group 4 farms. The average number of full-time paid workers is four per farm, with 40% of the farms employing two workers and 20% five of them. On the average, there are two full-time paid workers per 100 hectares. The average number of part-time workers is three per farm in Group 1, four in Group 2, thirteen in Group 3, and eight in Group 4.

The analyses indicate that small farms are much more labor intensive than large farms (Table 6). The number of work days per hectare per year is 86.1 for the smallest farms and 12.6 for the largest. The distribution of work days among the different manpower categories also points out the predominance of household labor. For all the farms in the study, household labor provides 61% of agricultural labor. On the other hand, there is a progressive decrease in household labor from Group 1 (the smallest farms) to Group 4 (the largest). By including the farmer as available manpower, this rate ranges from 97% in Groups 1 and 2 to 39% in Group 4.

Table 6. Distribution of Labor by Farm Group (Annual).

		Group 1	Group 2	Group 3	Group 4	Total
Family labor	Number of days	3,735	5,860	2,130	5,850	17,575
	Percentage	96.2	97.3	59.9	38.8	61.4
Part-time hired labor	Number of days	148	164	1,068	2,331	3,711
	Percentage	3.8	2.7	30.0	15.5	13.0
Full-time hired labor	Number of days	0	0	360	6,900	7,260
	Percentage	0	0	10.1	45.8	25.4
Total hired labor	Number of days	148	164	1,428	9,231	10,971
	Percentage	3.8	2.7	40.1	61.2	38.37
Total labor	Number of days	3,883	6,024	3,558	15,081	28,546
	Percentage	13.6	21.1	12.5	52.8	100
Number of hectares		45.1	151.6	181.3	1,194	1,571
Workdays/ha/yr		86.1	39.7	19.6	12.6	18.2

Agricultural Mechanization in the Study Zone

As the region of this case study is predominantly a cereal-growing area, the use of machinery has become relatively common among all farmers over the past fifteen years. Before 1960, only 23% of the farmers in Groups 2, 3 and 4 used tractors or combines. It was not until the 1970s that the use of tractors became common on small farms. Combines began to appear on these farms at a much later date than tractors. In fact, 57% of the farmers in Group 1 did not begin to use combines until the 1976–1980 period. The use of farm machinery throughout the region can be linked to the policy of incentives offered to farmers for mechanized farm equipment, a policy actively promoted during the 1970s.

Nowadays, the vast majority of sample farmers use machinery on their farms. Of the twenty-six farmers interviewed for this study, only one used neither tractor nor combine because of inaccessible land; another did not use a combine because he had plenty of household labor. These two farmers belonged to Group 2.

As stated previously, the ownership of tractors is limited to farmers in Groups 3 and 4. The other farmers (58% of the total) rent tractors. Even farmers who own tractors rent machinery (all of the farmers in Group 3 and 85% of them in Group 4). These farmers say that tractors from rental agencies are more powerful, and do the job faster and better than their own.

The ownership of combines is even more limited, as only 43% of the farmers in Group 4 own this type of machinery (11% of the total number of farmers).

Generally speaking, farmers first rent tractors and combines from other farmers, most often immediate neighbors. In the second place, they rent agricultural equipment from a government agency, SONAM (Societe Nationale de Motoculture). Farmers in Groups 1 and 2 (the small farms) rent their equipment almost exclusively from other farmers. On the contrary, farmers from Groups 3 and 4 (the large farms) prefer to rent from SONAM, from which they obtain all the tractors they need and 50% of the required combines. They also rent combines from other farmers. The average price for renting agricultural machinery is 6.4 TND per hour for each tractor and 21.5 TND per hour for each combine.

The use of rented equipment, tractors in particular, occurs most frequently in Group 3; the number of hours of rented tractor time totals 63 per 100 hectares. On the contrary, Group 4 farms use rented tractors the least. For Groups 1 and 2, the number of hours of rented tractor time is respectively 30 hours and 40 hours per 100 hectares.

Almost all sample farmers use mechanized plowing techniques, both for deep plowing and the plowing of stubble. Only in Groups 1 and 2 do a few farmers use draft animals either for deep plowing (17% in Group 2) or for superficial plowing (20% of the farmers in Groups 1 and 2). On the other hand, a majority of farmers in Groups 1 and 2 (86% and 62.5% respectively) plant by hand. In Group 3, most farmers use manual labor for planting;

in Group 4, all farmers use mechanical seeders. Fertilizers are also applied manually by all farmers in Group 1 and by 75% of the farmers in Group 2. In Group 3, 25% of the farmers spread fertilizer by hand. On Group 4 farms, this task is mechanized. Harvesting is not mechanized on 20% of the farms in Groups 1 and 2.

Determining Factors of Mechanization

Various factors influence the use of farm machinery by the farmers in our sample. These factors are linked, on the one hand, to production and farming systems and, on the other, to the political and socioeconomic environment.

We have already mentioned that the use of farm machinery has become fairly common in the research area over the past fifteen years, regardless of the size of the farm. Nonetheless, the extent to which it is used varies greatly according to farm size. On small farms, the use of machinery is most often limited to plowing (especially deep plowing, as transverse superficial plowing has been reduced to a minimum) and harvesting. On the large farms, all crop production operations are mechanized, including planting and applying fertilizers and herbicides. The minimal use of farm machinery on small farms is due to the fact that farmers do not own their own machinery and their capacity to rent it is very limited.

Farm size and, more precisely, privately owned land, also determine whether a farmer can own farm machinery. In the research area, 87% of the landowners own at least 40 hectares, which corresponds to the amount of land required to be eligible for a medium-term bank loan. Because owning a minimum number of hectares is a requirement for the acquisition and ownership of farm machinery, a close relationship also exists between the ownership of farm machinery and sharecropping or renting land.

As we have seen, farmers in the research area, regardless of farm size, use farm machinery for most operations in the production of cereal crops (plowing, harvesting and, to a lesser extent, planting). The essential operations related to forage crops are also mechanized, particularly plowing (which is entirely by machine) and harvesting (87% of which is mechanized). In spite of its relative importance, mechanization in forage production is less developed than in cereals production. For example, 67% of the farmers plant forage crops by hand. The use of machinery is much less intensive for food legumes. Other than plowing, which 66% of the farmers have mechanized, the remaining operations (planting, applying fertilizers and herbicides, harvesting) are manual on all farms. Legumes in Tunisia are still essentially a labor-intensive crop. They are also more frequently cultivated on small and medium-sized farms, where most of the labor comes from the household.

The close relationship between farm labor and type of crop was clearly revealed in the answers farmers gave to the following question: 'Do you think that the use of farm machinery is necessary, rather necessary, or unnecessary for the following tasks?' All farmers agreed upon the necessity of mechanization for plowing and harvesting cereal crops. A majority of farmers also stated that planting and the application of fertilizers should be mechanized (70% and 56%). On the other hand, for legumes 75% of the farmers agreed that it was necessary to mechanize the plowing, but more than 90% of them considered mechanization unnecessary for other operations (planting, applying fertilizers, harvesting).

For cereals, the need for mechanization can be explained essentially from a technical point of view. Thus, 80% of the farmers mentioned technical reasons to justify the use of farm machinery for cereal crops, while only 4% of them referred to the problems of manpower. For all farmers, the mechanization of cereal crops seemed to be an unavoidable development. None of the farmers could foresee any return to the use of draft animals or manual harvesting. This holds true in spite of the fact that there is often underemployment among the available household labor on small farms. Moreover, for these farms, the cost of renting farm machinery is prohibitive.

Socioeconomic and Political Environment

It would appear that widespread mechanization of farm labor in the research zone results from the environment in which the farms operate and an agricultural policy that has greatly encouraged the acquisition of farm machinery since the early 1970s by providing easier access to credit facilities. The effects of this policy have not only been felt among farmers who were thus able to purchase tractors and combines. All farmers can now rent farm machinery.

The practice of renting out farm machinery by farmers with large farms was favored by the fact that farmers with small farms were unable to apply for medium-term credit because they could not meet the basic security guarantee (owning at least forty hectares of land). This selective credit policy has resulted in a concentration of farm machinery on large and medium-sized privately-owned farms as well as on state-owned and cooperative farms. In addition to private owners and rental agencies, SONAM also rents out farm machinery. But as its fleet of equipment is limited, SONAM cannot satisfy the demand, a particular problem in bumper harvest years.

In addition, even farmers with large farms who own farm machinery use SONAM equipment. Farmers do so because their own equipment is insufficient during peak periods, they lack certain machines such as seeders and balers, they have no qualified tractor operators, and the more powerful SONAM equipment allows for the more rapid execution of farm operations. SONAM itself prefers to rent its machinery to medium-sized and large farms because the operations are performed more easily and farmers can pay in cash. SONAM refuses to let its equipment operate on certain farms when access is difficult or the land rough or hilly. It can also refuse its services if there is not a minimum number of work hours involved. SONAM also rents out its equipment at higher prices to private owners, charging an average 6.8 TND per hour for a tractor and 22.5 TND per hour for a combine.

For these reasons, farmers with small farms usually rent tractors and combines from private owners, a practice confirmed by our research data for the vast majority of farmers in Groups 1 and 2. In many cases, they rent from their neighbors who provide favorable conditions for payment. On the other hand, all farmers in Groups 3 and 4 rely upon SONAM. However, because SONAM doesn't have enough combines, some farmers in these groups (50%) also rent from private owners. In general, farmers who have medium-sized farms rent out their tractors, and rental agencies or farmers with large farms rent out combines. Farmers with medium-sized farms rent out their machinery for profit because this equipment requires a large capital investment and is often underused on their own farms. (In Group 3, all farmers consider their equipment to be underused while 57% of the farmers in Group 4 believe that their equipment is overused.) We must also emphasize that the growth of farm mechanization has led some farmers, as well as persons outside the agricultural sector, to purchase farm machinery and rent it out as a source of income.

Nevertheless, some farmers have major problems because they cannot obtain farm machinery when they need it, especially during peak periods (plowing and harvesting). Many farmers (79%) in the survey claimed that they did not have access to machinery at the precise time they wanted. This can be explained by the lack of available equipment for rent by private services or (especially) by SONAM. At the national level, SONAM at present has 350 tractors, which represent 1/100 of the national total. In the research area, there are four SONAM centers with 100 tractors and 15 combines, or one-fourth of all tractors and 15% of all combines in the region. The problems of availability of machinery affect farmers who must face long delays in farm operations such as plowing, planting, and harvesting. On some farms, planting occurs later than usual because farmers didn't plow in time. The unavailability of machinery accounts for the reduction in the amount of superficial plowing. In a good harvest year, competition for combines can result in large crop losses because of long delays in harvesting operations.

Sociocultural Factors

The growth of farm mechanization would also appear to result from an evolution in the sociocultural models of farmers. This evolution is closely linked to the determining influence of colonial farmers in the region. In their model of agriculture, mechanization played an essential role. This model remains omnipresent and, as a result, no farmer foresees any possible return to the use of draft animals even if it means a reduction in the expenses of farm mechanization (especially high for small farms, absorbing a large part of their income) and even if there is sufficient household labor available on the farm. In economic terms, mechanization for the farmer whose income still remains low could be considered an aberration.

The mechanization of farm operations on small farms also seems to result from a particular negative attitude toward manual labor, which has led to a refusal to perform time-consuming, arduous farm tasks by hand. This attitude has probably been accentuated by the fact that cereals cultivation in these regions is largely subject to the hazards of climate and provides only irregular, insignificant levels of income.

Socioeconomic Changes in Regional Agriculture Due to Mechanization

The mechanization of farm operations has brought about major changes in the way farms function. One of the most important changes is the growth in part-time farming. As farmers can now complete farm operations in much less time, mechanization has enabled them and their families to become involved in off-farm activities. On smaller farms, a large number of the farmers (50% in Group 2) hold off-farm jobs. For the most part, they are full- or part-time workers either in agriculture, road construction, or factories. They also open small shops or trade in livestock.

The increase in off-farm activities among farmers influences in many ways the manner in which their farms function. In fact, off-farm activity either reduces the function of the farm to one of mere food provision, or it serves as a means of intensifying farm production. In the first case, farm produce is exclusively for household consumption and the income from off-farm jobs finances daily family expenses. In the second case, the production strategy dominates and off-farm income is used to develop the farm.

The mechanization of farm operations is also a determining factor in whether or not farmers expand the areas they farm. For small farmers, the necessity of mechanizing farm operations and the heavy expenses involved limit any opportunities to cultivate more land. Because of the high costs of mechanization in particular, most small farmers work only the land that they own. As we have seen, most farmers in Group 1 (more than 70%) cultivate only their own land. Farmers who own medium-sized farms cannot enlarge their cultivated areas through sharecropping or renting due to the problems in acquiring farm machinery. In fact, when the area under cultivation exceeds thirty hectares, owning a tractor becomes indispensable because rental costs are prohibitive and dependence on service companies to perform farm operations too great. According to our data, the majority of farmers (88%) in Groups 2, 3, and 4 believe that owning a tractor is essential in order to organize their work better and greatly reduce expenses for renting machinery. As soon as farmers acquire a tractor, they often extend cultivated areas through renting or sharecropping.

In Group 4 farms, holdings rented from others constitute the largest part of the total area under cultivation. According to our data, 70% of the farmers in Group 4 increased their cultivated areas after having purchased tractors. On the average, land under cultivation increased by 50%. In Group 3, this phenomenon was less remarkable. In this group, the profitability of tractors appeared to be linked to renting them out to other farmers.

The effects of farm mechanization on the organization of crop production vary according to the type of farm. On some farms, there may be a diversification and intensification of farming. On others, mechanization encourages the development of single-crop farming. For example, small part-time farmers concentrate on raising cereal crops for household consumption and, in some cases, forage crops. On small and medium-sized farms, the mechanization of cereal crops frees household labor for other activities, which frequently leads to the diversification of farming. On the medium-sized farms in particular, intensive mechanization of cereal crops occurs along with the development of labor-intensive crops such as legumes. The diversification of farming also involves tree crops, forage crops, and livestock raising. If we consider the amount of tractor time hired per 100 hectares as a criterion for the intensity of mechanization, we note that this figure increases from Group 1 through Group 3 (from 30 hours to 63 hours per 100 hectares) at the same rate as the number of farmers who most diversify their production increases. In fact, the percentage of farmers who combine livestock raising, tree crops, large-scale crops, and legumes increases from 47% in Group 1 to 75% in Group 4. In Group 3, the level of mechanization matches the diversification of production. If we compare Group 3 with Group 4, we notice that the number of horsepower per 100 hectares is higher (149 HP as compared to 69 HP). Nevertheless, overmechanization on farms in Group 3 is offset by the cultivation of more labor-intensive crops such as legumes. These are grown on 7.2% of the cultivated area in Group 3 as compared to 2.8% in Group 4.

On Group 4 farms, the importance of mechanization is reflected in the predominance of totally mechanized crops (cereals and, to a lesser extent, forage crops). These crops are grown on almost 70% of the arable land (excluding fallow) as compared to 53% on farms in Group 3. On the other hand, very little land is used for cultivating labor-intensive crops (legumes). Moreover, 39% of the farmers in Group 4 have abandoned legume crops because of manpower problems. The importance of mechanization on large farms corresponds, therefore, to a low diversification of production.

If we examine the relationship between mechanization and the use of chemical fertilizers, we notice that the introduction of fertilizers followed the introduction of tractors. Before 1970, 80% of the farmers using tractors also used chemical fertilizers. In Group 1, the majority of farmers began to use tractors and phosphorus fertilizer only after 1976. In Groups 3 and 4, the majority of farmers adopted mechanization and fertilizers before 1970.

Manpower and the Division of Labor on Farms

There are relationships between the degree of farm mechanization and the status of labor (hired or household), on the one hand, and intensity in the use of labor, on the other. For example, it seems that the higher the rate of household labor, the lower the hours of tractor use per hectare. Thus, the number of household work days per year per hectare is 5 in Group 4 (the largest farms), but is 83 in Group 1 (the smallest).

As for hired labor, the number of days contributed by wage earners increases from 2 days per year per hectare in Groups 1 and 2 to 8 days in Groups 3 and 4. Moreover, the intensity of the use of labor is, in general, inversely proportional to the size of the farm and to the degree of mechanization. The total number of work days per hectare per year reaches 86 in Group 1, but is only 13 in Group 4.

On the average, 50% of the farmers in Group 3 have increased their deployment of household labor. This increase, which is also observed in the amount of hired labor, can be explained by the increase in cultivated land areas and by the introduction of new farming activities. This relationship confirms our previous observation, namely, that an increase in the use of labor is directly linked to the diversification of farming.

Even at the level of Group 4 farms, the increase in cultivated land area and the introduction of new crops has led to an increase in the number of wage-earning laborers. An increase in the use of manpower appears to be linked both to the type of crop and the type of farm operation.

While all farm operations related to cereal crops are mechanized, most operations for food legumes rely on manual labor. This is apparent from the number of work days necessary for the cultivation of cereals and legumes. Analyses of data from farms in Group 4 show that the number of work days for hired labor was 2.3 days per hectare per year for legumes, while it was only 0.7 days per hectare per year for cereals. Whether or not farm operations are mechanized greatly influences both the increased use of manpower and its status (full-time or part-time).

We can best observe the seasonal nature of farm employment on the large farms where cereal crops are grown on most of the available land and farm operations are entirely mechanized. Hiring occurs mainly just before the peak periods of plowing, planting, and harvesting. The seasonal nature of employment due to mechanization explains to some extent the problems that owners of large farms in particular have in hiring labor at the right moment in time. In fact, 71% of the farmers in Group 4 mentioned problems finding workers especially during the harvest season.

The mechanization of farm operations has also brought about a change in the distribution of tasks within the household, especially in the division of labor by gender. We can observe this change by comparing the distribution of work for one single crop in two farm groups that differ according to degree of mechanization, and by comparing the distribution of work for different crops within a single farm group. For example, on Group 1 farms, all members of the family (including wives and daughters) participate in certain of the manual labor tasks for cereals cultivation, such as applying fertilizer and collecting straw. However, other tasks, such as plowing and planting, even if performed manually, are always the responsibility of men. When we turn to the farms in Group 3, with the most advanced level of mechanization for cereal crops, we find that women have disappeared altogether from the work force. The operations, whether plowing, planting, applying fertilizer, or harvesting, are exclusively the preserve of the farmer and his sons. On the other hand, in this same group, most of the operations for the cultivation of legumes are manual, and the participation of women is especially important, except in the case of plowing, which is completely mechanized. Returning to Group 1 farms, we see that both the farmer and his wife perform the manual plowing for legumes.

Farm Relationships and Economic Institutions

Greater mechanization has stimulated the development of new farm relationships in the agricultural sector. In particular, farmers with small and medium-sized holdings have become almost totally dependent on the owners of machinery. We have seen that all farmers use farm machinery, almost 60% of them renting it from other farmers, from SONAM, or from rental agencies. Farmers who need to rent equipment are in a doubly dependent position vis-a-vis the owners. On the one hand, farmers pay for the use of tractors with a large part of their surplus production. On the other, farmers are at the mercy of machine owners for the timing of farm operations, a relationship that can seriously compromise the technical functioning and economic outcomes of the farms. This dependency at the level of work organization can have particularly serious repercussions on farmers who cultivate medium-sized farms and own no farm machinery.

The mechanization of farm operations also favors the development of dependent work relationships between small and large farmers. Small farmers often work on the farms of the latter and benefit from reduced rates and favorable payment arrangements for renting tractors, a situation that reinforces the relationship of dependency.

Thus, the mechanization of farm operations has resulted in stronger economic control by farmers who own large farms. More recently, owners of medium-sized farms have been able to acquire machinery, as have other persons not employed in agriculture. For these individuals, renting out this equipment is an important source of income and a way of profiting from machinery that is underused on their farms. Farmers who work large farms are less interested in providing this service because of the large areas they cultivate and the problems related to renting out their machinery, such as damaged machinery and unavailable drivers.

The mechanization of farm operations has also resulted in the development of relationships between farmers and credit institutions and agricultural development organizations. Bank loans helped many farmers in Group 4 to acquire farm machinery. Moreover, the increased use of farm machinery by the owners of small and medium-sized farms has encouraged the latter to join agricultural development projects that provide farm machine rentals on credit (for example, the APMANE project in northeastern Tunisia), as well as short-term loans in kind such as fertilizer and seeds.

As we have stated, the process of farm mechanization has led to the increased use of improved seed, chemical fertilizers, and herbicides. A large proportion of farmers (46% of all farmers in the study and 86% of Group 4 farmers) have applied for short-term loans to purchase seed and fertilizer. In regard to cereals, the increased use of new crop techniques in general has resulted in closer relationships between farmers in the various groups and the Cereals Board, which plays an ever-increasing role in providing improved seed and chemical fertilizer to farmers as well as marketing their harvests.

Farm Viability

The trend toward greater mechanization has not brought about an overall increase in crop yields for all farms. We note a rather large difference between the average yields for durum wheat on the sample farms. Between Group 1 (the smallest farms) and Group 4 (the largest) there is an 85% difference in yield, namely from 980 to 1790 kilograms per hectare. In Groups 2 and 3, the durum wheat yields are 1120 kg/ha and 1230 kg/ha respectively. Mechanization does not necessarily lead to high yields in the absence of other factors.

Moreover, mechanization on small farms is often inadequate. It is frequently limited to deep plowing and insufficient harrowing, which may not occur at the most opportune times due to problems related to the availability of machinery.

There are other factors that determine the level of output on Group 1 farms, notably the use of improved seed and fertilizer. As we have already mentioned, many farmers rarely exchange their seed and 43% of Group 1 farmers never use either phosphorus or nitrate fertilizers. In addition, small farms are often located on terrain poorly suited to growing crops.

One of the main effects of mechanization on small farms is an increase in farm expenses. According to farmers in Groups 1 and 2, mechanization expenses for cereal crops represent more than 60% of the total. We should mention that for these farms household reproduction strategies hold priority over production strategies. The essential role of the farm is to meet part of the family's food needs, with all other needs met by income from off-farm activities. For this reason, in spite of very low production levels, the majority of these small farms continue to exist. Yet as soon as the logic of production prevails, small and medium-sized farms owning no farm machinery experience very serious problems.

These farms are also at very high risk because of uncertain weather conditions. These difficulties increase in proportion to the cultivated surface area, as costs for renting farm machinery have become too high. Most of the farmers in Group 2 (88%) state that their main problems are insufficient cultivated land surface and the high costs of mechanization. In addition to the costs of mechanization, we note an increase in other input expenses for Group 2. The major weakness of small and medium-sized farms, evident in the logic of production and vulnerability to weather conditions, is shown by the great difference in cultivated area from one year to another. Many farmers in this category, after a bad harvest year, give up the lands that they have either rented or sharecropped. High rental costs and problems in acquiring machinery (credit ineligibility) are the main reasons why these farms are blocked in their development.

The problems in acquiring tractors are made even more difficult not only because of their high cost (the purchase price of a tractor ranges between 10,000 and 15,000 TND depending on horsepower) but because farms under 30 hectares do not have enough surface area to profit from the use of machinery. Nonetheless, ownership of a tractor is essential for the development of these farms, especially as there are no cooperatives for farm machinery.

Farmers in Group 3 have discovered that farm machinery makes it possible to extend cultivated areas, reduce mechanization expenses, better organize operations, and create a new source of income from renting out the machinery. In general, mechanization has helped this group of farmers to conserve and consolidate their farms. For them, the combination of land and tractor ownership has become an essential source of profit in agriculture. Nevertheless, in spite of the ownership of tractors, the mechanization expenses for cereal crops remain very high in Group 3 because the farmers are obliged to rent combines. The overall economic performance of these farms is due, for the most part, to the greater diversification of farming operations as well as effective management that makes maximum use of household labor.

On Group 4 farms, mechanization expenses account for more than 50% of all expenses, according to farmers in our sample. For cereal and forage crops, input costs are also very high (30% of the total). Although in the research zone the average cereal yields of these large farms remain lower than those in major cereal-growing areas, the income of farmers can be relatively high due to the large areas they cultivate. Up until now, because they have had access to farm machinery for a long period, they have been able to expand their cultivated areas. However, the tremendous increase in the costs of farm machinery makes the replacement of aging equipment very difficult and would cause very serious problems for the management of this type of farm. An agricultural services official in the research area indicated to us during the survey that even the owners of large farms hesitate to replace their machinery because of the costs. The result at the local level is that there are few available machines and many farmers who seek to rent them, particularly from SONAM.

Conclusions

The mechanization of farm operations brought about major changes in agriculture in the research area. These different changes (technological, economic, social) affected farmers in various ways depending upon the size of their farms and their crop production systems. Generally speaking, mechanization resulted in the development of new social and economic relationships in the sphere of farm production. Similarly, the agricultural sector became more closely connected with other sectors of economic activity. We can mention, in particular, the increased use of manufactured goods in farm households.

If the introduction of farm mechanization to small farmers caused a major increase in production costs, it nonetheless did not threaten their existence. When their incomes decreased as a result of higher costs, small farmers looked for new sources of income outside of agriculture. We must also mention that there is a considerable underemployment of household labor on these farms. On medium-sized farms, mechanization has made long-term farm viability more difficult.

Mechanization has had the most positive effects on the large farms because it has made possible the exploitation of much larger surface areas. However, it has also resulted in extensive farming systems where land left in fallow is still very important and where there has been a major decrease in the level of employment.

The problems resulting from farm mechanization bring up the question of the applicability of a single technological type for farms whose structures and systems are so diverse. In the research area at present, farm mechanization has become an irreversible process closely linked to the predominance of large-scale crops, especially cereals.

To counteract the negative effects of farm mechanization, in regard to both production costs and employment, the solution would appear to be a diversification of the farming practiced in this region, essentially in the development of food legume crops, fruit trees, and livestock raising. This would result not only in higher incomes for small and medium-sized farms in particular but also in a higher return for household labor on these farms.

In addition, a reduction in mechanization costs, in particular for owners of small and medium-sized farms, would stem from easier access to medium-term loans. However, for many of these farmers, the costs of individually purchasing and using machinery on a limited amount of land would be too expensive and unprofitable. Therefore, a more effective solution would be the establishment of cooperatives for farm machinery. There should also be more machinery and rental services (such as SONAM) available for farmers.

Finally, for the longer term, a local industry for farm machinery is vitally necessary to solve problems such as the high costs of imported equipment and the shortage of spare parts.

References

INS (Institut National des Statistiques). 1984. Recensement de la population 1956, 1966, et 1980. Tunis: Institut National des Statistiques.

MA (Ministère de l'Agriculture). 1981. Enquête agricole de base 1976 et 1980. Tunis: Direction de la Planification, des Statistiques et des Analyses Economiques, Ministére de l'Agriculture.

MA (Ministère de l'Agriculture). 1985. Enquête agricole de base 1985 et 1980. Tunis: Direction de la Planification, des Statistiques et des Analyses Economiques, Ministére de l'Agriculture.

SEA (Secretariat d'Etat a l'Agriculture). n.d. Structures des exploitations agricoles, enquête 1961–1962. Tunis: Secretariat d'Etat a l'Agriculture.